Engineering as a Career

by
Kenneth G. Budinski
Technical Director,
Bud Labs USA

9 781300 676546

Dedication

This book is dedicated to Steven T. Budinski
A good son and a great engineer

Acknowledgments

This book was made possible by Lynne's typing, Ann Marie's art work and Mark's ability to put everything together and to make computers do their job. Thank you also to ASM International, John VanDursen, and the Kodak Engineering Retirees group for their helpful reviews. We also acknowledge the testimonials on different types of engineering from Jane Mulkern, Carole Mahady, Dr. Christine Landry, Michael Budinski, George Plint and John Budinski. We also acknowledge the prefatory comments on engineering as a career from Dr. Bharat Bhushan a world renowned engineering educator and researcher. The photographs in this book were either taken by the author or came from his photo collection.

Content

Chapter 1   Career Exploration   Page 1

What is a career?

The shopping process

Self-evaluation

Summary

Chapter 2   Engineering Terms   Page 15
and Definitions

Engineering jargon

Summary

Chapter 3   What Engineers Do?   Page 37

Working during school

Your first job

Having a mentor

Design engineering

Research and development

Career advancement

Summary

Chapter 4   Important Engineering   Page 61
          Achievements

Prehistoric times

Early civilizations

 Greek/Roman/Asian civilizations

The medieval era

The industrial revolution

Summary

Chapter 5   Do your Interests Match   Page 83
          an Engineering Career

What makes you who you are?

Physical characteristics

Personality traits

Beliefs

Summary

Chapter 6   Schooling Requirements   Page 97
          for an Engineering Career

Preschool for engineering

Grammar school

High School

Undergraduate Engineering school

Graduate school

Choosing a school

Accreditation

Summary

Chapter 7   Types of Engineering   Page 115

Choices

Electrical Engineering

Mechanical engineering

Computer engineering

Civil engineering

Chemical engineering

Industrial engineering

Materials/metallurgical engineering

Mining engineering

Aerospace engineering

Biomechanical/biomedical engineering

Petroleum engineering

Summary

Chapter 8   Career Aids   Page 157

Orientation

Technical Societies

Continuing education

Networking

Summary

Chapter 9   Life as an Engineer   Page 171

Start of career

Mid-career

Career maturity

Career summary/retirement

Chapter 10   Future Engineering   Page 179
          Challenges

Environmental issues

Economic issues

Society issues

Epilogue   Page 191

Appendix   Page 197

Preface

This book is a response to a perceived shortage of engineers in the USA. The 2019 to 2021 pandemic shut down all of the engineering conferences that I regularly went to and all that I could see throughout the pandemic were engineering needs. For example, the mask requirements, air filtration, and virus testing requirements were significant engineering challenges.

I am writing this in fall of 2022 and the pandemic has been diminished in some countries mostly by vaccination, but there are still countless engineering challenges to overcome issues of living with deadly viruses in the air. The environmental challenges that are concurrent with the pandemic also cry out for more engineering assistance. The energy issues that exist cannot be solved without lots of good engineers working on the processes and equipment that will be needed to transform the world into one that can accommodate our incredible sustainability issues.

This book is an attempt to interest young people in engineering as a career. It is a narrative by a still-practicing engineer about what it takes to become and engineer, what an engineer's life may look like, and what might be the product of a person's life spent in the engineering field. The book starts with an explanation of engineering as a career, how it compares with other candidate careers, and young readers are encouraged to use a selection matrix to do career comparisons, to put some numbers on candidate choices.

The second chapter defines the terms that apply to engineering so that readers are not swamped by the jargon of the field, and there is lots of jargon, but not as much as the medical fields. They win. Chapter 3 chronicles what engineers do on a daily basis. Of course, what an engineer does, depends on who they work for, but there are basic things like design, that are part of most engineering jobs. Chapter 4 reviews some transformational engineering achievements that have allowed civilizations to progress from hunting and gathering to what we have today.

Chapter 5 prompts readers to compare their personal interests, personality, abilities, and strengths with those that typify engineering. Chapter 6 outlines typical schooling requirements for an engineering career. Chapter 7 describes the more popular types of engineering, based upon degree. Chapter 7 also includes a number of testimonials from various types of engineers. They answer the question why they chose a certain type of engineering for their career. One of the most famous engineers in the world answers the question: "Why I choose engineering as a career?" following this preface. Chapter 8 discusses "career aids"; these include things like networking, participation in technical societies, and opportunities for continued learning in engineering. Chapter 9 describes what a life in engineering would look like. What would your early career look like, what would mid-career look like. The last chapter predicts the challenges that lie ahead in engineering. What might engineering look like in the future?

This book is an authentic account of an engineer's life in the USA during its manufacturing era. Things were easy simply because most engineers worked in large companies and jobs were often for life. In 2022, many of the large manufacturing companies are gone, but the engineering challenges are much greater than those that existed during the manufacturing era. The stakes are higher; solving the world's environmental, energy, food, and housing problems will require a level of engineering expertise greater than at any time in the past. Young people are needed in engineering more than any time in history. Are you willing to sign up to solve our existential problems? This book hopefully will guide you through the decision-making process.

kgb

Why I choose engineering as a career?

Prof. Bharat Bhushan, Academy Professor (The Ohio State University, San Jose, CA)

In my high school, I took Physics, Chemistry, Mathematics and Biology in addition to literature and social sciences. I loved physics, chemistry and math and did not care for biology. In India, preference has been to go to either engineering or medicine as these professions are highly paid. Since I did not like biology and was a gifted student in physics, chemistry and math, I chose engineering as a career. I liked working with machines and materials, so I chose mechanical engineering. I received a BS (Hons) in Mechanical engineering in May 1970 from Birla Institute of Technology and Science (BITS) Pilani, India.

During my BS, I had a preference in production engineering and reliability. Therefore, I chose to pursue graduate research in Tribology at Massachusetts Institute of Technology (MIT), Cambridge, USA. I finished a MS degree in one year. I presented my first paper on Sliding Surface Interface Temperatures at ASME/STLE Lubrication Conference, New York in October 1972, which was published in ASME Journal of Lubrication Technology in 1973. After that I continued in the field and have worked both in corporate research labs and academia. I have spent 53 years in the Tribology profession and cannot slow down.

I have loved tribology research to develop fundamental understanding and solve engineering problems. I cannot wait to get up and go to work. I have been fortunate to work in a field I love. Work is not work if you love what you do. It is not just to collect a paycheck. I am the lucky one. My message is to chase a profession you love, and you will excel and lead a satisfactory life.

Dr. Bharat Bhushan is an Academy Professor (San Jose, CA), and has served as Ohio Eminent Scholar and Howard D. Winbigler Professor, and Director of Nanoprobe Laboratory for Bio-& Nanotechnology and Biomimetics at Ohio State University, Columbus, Ohio. He holds a B.S., two M.S., a Ph.D in mechanical engineering, MBA, and five honorary doctorates, a total of 10 College degrees. His research interests include Fundamental studies in the interdisciplinary areas of Bio/nanotribology/nanomechanics, Nanomaterials Characterization, Scanning Probe Techniques, Magnetic Storage, Bio/nanotechnology, Nanomanufactutring, Bioinspired Liquid Repellency, Self-cleaning, Anti-icing, Anti-fouling, and water Harvesting, Science and Technology Policy. He has authored 10 books and 900 plus papers and holds 25 US and foreign patents. He has previously worked for Mechanical Technology Inc., SKF and IBM Almaden Research Center, San Jose, CA.

Chapter 1 Career Exploration

What is engineering?

I do not recall when I first heard the term "engineer". Maybe I heard it in the movies as the person who drives a train, or maintains an apartment building (building engineer). However, these jobs are not included in the "engineering" that this book is about. The next chapter contains many terms and definitions pertaining to engineering, but let us start by defining engineering:

Engineering, n. The profession composed of people educated to employ science, research, existing truths, analytical techniques, and intuition to conceive, design, develop, and oversee the production of things and processes that solve problems, meet needs and wants to make civilization possible.

Engineers are people who practice engineering. The world has always had engineers, people who developed useful things that solved a problem or made something easier. Archeological studies show how inventions evolved to allow civilizations to exist. The person or persons who invented fire-on-demand, pottery, stone tools, shoes, etc. were engineers before there was a name for what they did. Certain people were given gifts that made them an innovator, a creator of new useful things. We have a chapter on transformative inventions,

1

inventions that forever transformed the world. For example, recent archeological digs produced an intact saddle for a horse that somebody made in about the year 500BC. The saddle and the bridle for horses were transformative inventions. They forever made horses easier to use for work, transportation, and for pleasure. Both are still in use with basically the same design that goes back more than 2000 years. People who made the sorts of things that made life easier were the original engineers.

The schools and Universities that teach engineering as we know it did not exist until about the 19th century. By then, the world, through research conducted by many exceptional people, knew enough of the fundamentals of nature to allow the teaching of engineering.. People could go to school to learn how to conceive designs, how to analyze designs to see if they would work without building them and they developed means for building new things. The fundamentals of nature that serve as the basis for engineering are sometimes called first principles. They are truths that always apply and they allow things to be made and to work as intended.

Mathematics as a field of study goes back to the start of recorded history. People in early civilizations started counting things like ears of corn produced from a garden; then when money was invented, mathematical techniques were developed to keep track of money and to manipulate money. Then some ancient "thinkers' invented geometry, then trigonometry, then calculus, and by the Middle Ages mathematics and the languages to use them were well developed. First principles are rules that apply to nature. The ultimate development of a first principle is to make it a mathematical expression, an equation. For example, the mechanical force produced when a golf club strikes a golf ball is a product of the mass of the ball and the acceleration produced by the club speed: $F = M(mass)A(acceleration)$. All engineering students learn to love and respect and use this equation in design. It is a first principle equation.

By the Middle Ages, first principles were discovered in many areas: astrology (what the planets do), gravity, flow of liquids, transfer of heat and energy, properties of materials – many important issues that pertain to making and doing things. All sorts of first principles were put into mathematical equations and these are the types of things that are taught in engineering school.

This book has a chapter on specifics of what is taught in engineering school. However, an engineer is a person who uses the first principles taught in engineering school, experiential knowledge, and intuition to conceive and build the things and processes that civilizations need. The overarching goal of this book is more young people choosing engineering as a career. We will also try to show what a typical engineering career looks like.

What is a Career?

A career is what you choose to do for your life's work, and it ideally financially supports you and a family. Some careers, like a football quarterback, only support the very few. My niece wanted a career as a professional dancer. She even got a college degree in dance, but she could not financially support herself dancing in local productions. Instead, she ended up getting an additional college degree in physical education and teaching dance for high school musical productions.

A career is what you want to do with your life, but being a creative dancer may not pay your bills.

A career is different from a job. My first job was delivering a once-per-week print newspaper. I started at 7 years old and delivered 300 newspapers for 65 cents. I did not want to make a career out of newspaper delivery.

I never thought about a career in grammar school, but because my parents both worked in nearby factories I was exposed to this work. I knew very little about professional careers until I reached high school. We had career talks on a regular basis from accountants, policeman, journalists, firemen, engineers, and doctors. It was from these talks that I started to form the concept of a career.

The way that my father described a career was straight-forward. At the time, my father was working for General Motors at the windshield-wiper motor factory. While working there as a machinist/toolmaker he encountered cooperative engineering students at GM's engineering school, General Motor Institute (GMI). At GMI, students worked six months of the year and went to school six months of the year. My father told my brother and I that we were to go to GMI to become engineers. I was assigned to the local GM carburetor factory in 1958. It was a five-year program and students worked full time on a thesis project in the fifth year.

Many of my high school classmates pursued careers based upon their parents' suggestion. One of my classmates had a father who owned a car dealership, and he made this his career. Three brothers who were my classmates in high school made a career of their father's plumbing business. Another classmate took over the family bakery

Making lots of money is not a good basis for selecting a career; enjoying your work is more important. The ultimate career is getting paid for doing something that you like to do. Very few can make a living playing professional sports or performing creative dance, but many can make a career out of something that they enjoy doing. When my middle son was in high school, he used to go into my basement workshop and turn metal on my lathe. He ended up making lathe work his career. He has been a tool maker specializing in precision lathe work for about 30 years.

The best career is one where you like what you are doing.

Evaluating Candidate Careers

One of the most incredible career choice aids that I witnessed was on Chinese television in the 1980s. I visited China as a guest of the Chinese Academy of Sciences in 1983 and our hosts would drop us at our hotel each night and I would watch television. The programs on TV were about how to make things. These

5

how-to shows were wonderful; I could see how particular goods needed by society were made. I watched shows about making bicycles, batteries, silverware, pots and pans, harvesting soybeans, and making taffy; the shows described how these things were made or manufactured in factories. This type of educational programming allows young people to learn about career choice. One could see what a baker, a bricklayer, and a steelworker did. I speculate now that this type of education programming on Chinese television played a role in the country's incredible manufacturing capabilities throughout history.

Some young people are born with skills and traits that point to a particular career; some are guided by parents, but most young people need to find a way to compare careers. In my materials engineering work, I often compare candidate materials for an application. A tool that I found helpful is what I call a decision matrix. It helps to quantify decisions. Here is how it works: put career candidates in a vertical column and selection properties in columns relating to career properties. Suppose you are considering a career as an engineer, accountant, lawyer, doctor, or teacher. The candidate careers are put in the first column. Then columns are established on career properties such as salary, type of work, challenge, travel, benefits, societal benefits, and regular workday. The goal is to list the five or ten most important properties offered by a lifetime career.

The numerical ratings for each property are your assessment from 1 to 10 (10 being best). You can weigh certain properties. For example, if you are close with your family you can double the numerical value that you assign to that column in a career-decision matrix. Figure 1 at the end of the chapter is an example of a decision matrix on engineering and competitive careers. A young person can do many iterations of a decision matrix, but the concept is to try and quantify a comparison of careers using career properties that are important to them.

Some people want a career that offers regular 9 a.m. to 5 p.m. work hours. If this is a desired career attribute item, careers like military police or nursing will rate low in that attribute. A career decision matrix offers a simple way to evaluate various careers for suitability. For example, along with likes and dislikes, you can also do a career matrix listing skills required rather than job attributes. In this case, instead of salary, benefits, or challenges you could list required skills like math, manual dexterity, ability to multitask, attention to detail, or report writing.

Careers should be shopped as one would comparison-shop an important purchase

Self-Evaluation

I recommend the use of a decision matrix, but a young person must also evaluate themselves. Ask yourself, what are your physical and mental limits? What are you good at? What do you like to do, but may not be good at?

Know thyself — Shakespeare

Limited ability is both a real and important thing to understand about yourself. For example, I do not have the patience to read directions and spend time learning how to assemble something. When I buy something that needs to be assembled, I just start assembling and see what happens. This may be foolhardy, but that is my personality. Personality traits are just as much a part of career selection as schooling. If you are an introvert and are uncomfortable speaking to groups, you may not want to pursue teaching or sales as a career choice. Young people need to consider their personality traits when exploring careers. Careers that require continual interface with the public belong to extroverts, while an introvert may be a great candidate for an accounting career. Thus, a helpful career guidance step is to establish a personal profile. Your profile should include:

- Your gifts—things that are easy for you, proficiencies, and physical advantages
- Your situation—family, siblings, income, where you live, or want to live
- Your interests—enjoy the outdoors, enjoy exercising, intense reader/studier, loves animals, enjoys being around people
- Your personality traits—patient, methodical, conscientious, or gregarious
- Your limitations

Selecting a career requires matching your personal profile to a career profile

Your personal profile will not only affect your career choice, but your ability to progress in a career. For example, mixing chemicals and studying results is one career path while managing the work of others may be another path.

Consider what career growth looks like

Differences between Science and Engineering

Science, technology, engineering, and math are referred to as STEM in the USA. Science describes the activity directed towards adding to the world's body of knowledge in a particular field. The field of science can range from astrophysics to xerography. Just about anything can be a scientific field—an area of education, study, research, or exploration. The pure sciences usually include chemistry, physics, and biology, which are also the science subjects taught in most schools. Mathematics is a necessary part of these basic sciences and is also taught in schools. Mathematics is an essential tool in all sciences, engineering, and technology.

Technology, in STEM, stands for the use and role of electronic devices and data manipulation. Companies that sell mobile devices are considered part of the "Technology (Tech) Sector. If something involves the use of a computer, it is part of the technology sector. We defined engineering in this book's prefatory remarks as the field of human endeavor that concerns design and building useful items to make life better. Engineers use the basic sciences to make useful items while scientists study something to increase the body of knowledge. My MS thesis was "The effect of solute concentration on the yield strength of Alpha iron." It was a study to find out what happens to the element iron when you refine it to the point where nearly all the carbon is removed. This study was a scientific study; however the results assisted

9

the improvement of engineering applications. Science probes fundamentals, while engineering applies what science has learned to make useful things.

Generally, companies that hire college graduates usually staff engineering departments with BS and MS graduates and their research department with MS and PhD graduates. People with PhD's in chemistry often work on understanding mechanisms concerning how the company's chemical products work; they look for new, useful chemicals. People with a PhD in physics may be employed by an energy company to research cleaner ways to produce energy, like desktop atomic reactors or a device that runs on the entropy of the universe. It usually takes an advanced degree beyond a BS to work in science, but it is the nature of the work that differentiates science from engineering. Engineers make useful products/processes; scientists study fundamentals to determine new concepts and mechanisms that may someday have utility.

When I worked in a large chemical manufacturing company the research scientists were studying ways to make chemicals change by the action of photon activity. They developed concepts for new photographic films and also discovered digital photography. The company engineers attempted to translate the science into something that could be used to make a profit. Thus, we had to find the way to put a new photon receptor chemical on a film support that could be manufactured to make a profit. Unfortunately, we failed to translate digital photography into a profitable product. Other engineers did this better at a later date.

As I write this, artificial intelligence (AI) has become a saleable product in the form of Chatbot (as one example), but I worked on the concepts of AI about 40 years ago (it was just science then). Thus, the difference between science and engineers may be in the form of time, in this case, decades. Science deals with concepts; engineering makes things that people need in their daily lives.

Summary

Traditionally, careers start for young people when they complete high school or college but young people may need career-related training and experiences beginning in secondary school. I began career planning with my sons when they started high school, but I tried to show them many different careers well before that through public events, travel, and sports.

Career exploration starts by becoming familiar with different careers and with yourself, your likes, dislikes, abilities, limitations, and who you are as an individual.

All careers have requirements, and certain attributes are needed for each. For example, when I was five or so years old, I wanted to be a doctor. However, after a few bicycle accidents as a child, I learned that I got faint at the sight of blood. This would not have been helpful in a career in medicine so I started thinking about other career choices.

A good way to determine your suitability for a career is to create a personal profile. List your native abilities, current life situation, likes, dislikes, personality traits, and limitations. Think about career candidates and list the career requirements. Match your personal profile with career requirements to arrive at candidate careers. (Later in this book, I will go over career requirements in engineering and illustrate attributes of the best candidate.

Profession	Job Attributes										Total Score	Ranking
	Good Salary	Meaningful work	High Demand in job Market	Stable Employment	Opportunity to travel	Requires working in teams	Can be individual contributor	Requires life long education	Can work from home	Hands on Work		
Accountant	6	5	8	8	2	1	9	3	9	7	58	4
Teacher	9	7	4	9	2	4	6	9	1	3	54	6
Engineer	9	10	10	10	10	10	4	8	5	7	83	1
Doctor	10	8	9	9	2	8	3	10	1	8	68	2
Lawyer	7	5	3	3	3	2	7	4	8	1	43	10
Artist/Musician	1	8	2	1	8	1	10	2	10	10	53	7
Computer Coder/etc.	6	2	8	2	3	1	5	7	9	6	49	8
Chemist	3	3	6	8	3	2	3	7	2	7	44	9
Nurse	5	8	9	9	2	5	3	8	1	8	58	5
Agronomist	8	10	3	9	1	1	9	7	9	9	66	3

Figure 1 *This is an illustration of the use of a decision matrix to explore careers. You decide on the careers that you want to compare, and the job attributes that you wish to use to evaluate careers. Then you assign a number from 1 to 10 with 10 being best to rate each attribute for each career candidate. The best career for you will have the highest score.*

Figure 2 People love their automobiles. There are scores of companies that make them and each company has a significant number of engineers designing the autos, the parts that go in them, and the machines that build them

Chapter 2 Engineering Terms and Definitions

Every career has associated jargon. Engineering, of course, is no exception. We are devoting a chapter to terms and definitions because a key part of career planning is becoming familiar with what things are called in a career. Just today I called a local hospital to get the name of an administrator and I ended up on the phone with the "utilization manager". Of course, I had to ask; what is a utilization manager in a major hospital. She said that they are responsible for dealing with insurance companies to pay for health services. Who knew? However, we all know that medical insurance is a key part of the medical profession worldwide. I now know that a utilization manager is a career in medical insurance. And that is also why terms and definitions are important in engineering. We will not go too deep, but we need to establish core terms and definitions.

We start with engineer and related terms. We will present a personal commentary when appropriate so this chapter is not like a glossary.

An engineer is a person with the knowledge and skills to conceive, design, and implement something that solves a problem or makes something better.

Most definitions of "engineer" included the words "science" and "math", but people have been conceiving designing and building useful things since the start of what we now call civilization. Prehistoric people made tools for building and hunting from sharpened rocks. The person who conceived of putting a sharpened rock on the end of a stick (to make the ax) could be considered to be what we now call an "engineer". Archeologists believe that the first "wheel" was used for making pottery from clay. Thus, the greatest invention of all time, the wheel, originally was not used for transportation. Another "engineer" conceived of putting two potter's wheels on an axle and invented a chariot. Thus, engineers did not have formal science and math to create things but science and math eventually explained why things work. The wheel works because rolling friction is lower than sliding friction. The axle works because it concentrates compressive stress along a line. Inventions from prehistoric engineers changed the world forever.

The internet states that the word "engineer" comes from the Latin word "ingenium" which means to contrive or devise. Whatever the derivation, the intended meaning in 2022 is a person who can create useful objects and solve problems. Neanderthals needed weapons to hunt for food. Archeologists uncovered sharpened sticks that they believed to be used for hunting. They also uncovered sticks with sharpened stone and arrowheads which could better penetrate a mastodon hide. The spear had been engineered. This invention is still in use after more than 100,000 years. Some people living in remote areas still hunt with spear and arrow.

Lots of inventions occurred in pharaonic Egypt. The Egyptians invented a written language as a way of documenting their lives so that future generations could see how they lived.

Figure 3 *The invention of the sail was one of the most important inventions of "early" engineers. Egyptian tombs show drawings of boats propelled by sails on the Nile River at least 5000 years ago. Sails are still used worldwide to propel all sorts of vessels. How is that for a great invention?*

The use of animals to pull farming tools and vehicles, bows and arrows, and, of course, magnificent pyramids were the products of Egyptian engineers. The engineering career has existed for at least 5000 years. Certain people developed the aptitude to conceive design and execute things that solved a problem or made things easier. Egyptians had positions of reverence for the "thinkers" who gave their civilization so many new things. In my engineering specialty, tribology, the art and science of interacting surfaces, the Egyptians gave the world the concept of lubrication. Petroglyphs on their monuments clearly show stone jars of liquid being poured in front of large stones being pulled on sleds by people. Chemical analysis of wooden axels on a chariot in an Egyptian tomb showed that they used animal fat and other ingredients to lubricate the chariot bearings.

How did civilizations like pharaonic Egypt have advanced technologies when much of the world was still living relatively primitively? Their "engineers" allowed their progress over others. Engineering was well established in some civilizations more than 5000 years ago.

The first college in the USA to grant degrees in engineering is reported to be the US Military Academy at WestPoint in 1817. One of my favorite engineering schools, Rensselaer Polytechnic Institute (RPI), in Troy, N.Y., followed with engineering degrees in 1825. My graduate school, Michigan Tech, granted engineering degrees in 1885. My undergraduate engineering school General Motors Institute (GMI), now Kettering University, established cooperative engineering programs where students worked in GM factories one half of the year and went to GMI the other half of the year. They graduated experienced engineers. It took a bit longer (five years), but both parties benefited. GM got experienced engineers; students paid for their college with their salaries from the working portion of the year. They started this in 1919 and it continues today.

In 2022 there were about 340 colleges and universities in the USA that offered engineering degrees and there were at least 900 worldwide. Thus, engineering has an incredible history of achievements. Besides the enormous Egyptian engineering feats, the Romans built roads, harbors, aqueducts, and arenas, some of these are still in use after 2000 years. The famous Leonardo Da Vinci was an incredible artist, but also an engineer extraordinaire. One time I viewed a display of some of his machine drawings at an ancient monastery in Milan. He invented all sorts of devices that are still used. His day job was to design military equipment for his patron, the Duke. He invented what we now know as the "tank". He invented the screw thread, gears and many mechanical components that are the bases of modern machines. Engineers allow civilization to exist.

"Engineering Technology" is a field of engineering that usually requires less formal education than "engineering".

Engineering Technology may not be a worldwide part of engineering, but in the USA, many state universities have community colleges that are smaller in size than university centers and many offer engineering technology degrees that can usually be completed in two years as opposed to the four years minimum for bachelor degrees in engineering. When I worked in the chemical process industry, people with two-year engineering technology degrees often worked as technicians.

Technician, n., a person who works under the supervision of another with more education

When I worked in a materials engineering laboratory, engineers would establish a project or study that involved a series of laboratory tests. Technicians would perform the laboratory tests and the project engineer would analyze the test results and establish the next project step.

Some engineering schools offer four-year engineering technology degrees that involved less rigorous course work than engineer degrees, but four-year engineering technology graduates may need to take additional courses to get into engineering graduate schools.

If an engineering school graduate in the USA is to work in a field that involves public safety, he or she may need to take a state test to become a professional engineer (PE).

Professional Engineer, n., in the USA a graduate of an accredited bachelor-degree engineering program who passed a series of state licensing tests.

If an engineer wants to work in building bridges, highways or similar public facilities he or she will need a PE license to meet most employer requirements. PE licenses are often required to testify on court cases that involve engineering matters.

It is common practice in many US states to offer "Engineer in training" (EIT) tests after college graduation. These tests are normally in basic engineering principles and the final PE license can only be obtained after five years of work as an engineer. Thus, there are tests after college to become an EIT and a final test after five years to obtain a PE license.

Another important aspect of a career in engineering is apprenticeship.

Apprentice n., a person who works under a person with established skills (master) long enough to acquire a certain minimum amount of skill in a field.

Apprentices probably go back to pharaonic Egypt. The massive stone shapes used in their structures required incredible skills in stone carving. In fact, every skill needs to be passed on through some formalized program like apprenticeships. In some states it may take the form of working for a master for four years and doing several thousand hours of related work and study.

Apprenticeships are strongly related to engineering. It is the engineer's job to conceive, design and build something. However, the design and build phase requires help from others. When I worked in machine design, I would sketch a machine concept; then make a full-size assembly drawing of the machine, then have drafts people make the detail drawing of parts. The "details" were made in the machine shop.

Apprentices usually rough shaped metal parts working under a master machinist. Thus, engineers do not do all of the design, drafters do the detail drawing and engineers do not machine the parts needed to build something. Machinists and apprentices fabricate the parts needed to build a machine.

Figure 4 *Surface grinders in a typical machine shop. Machine shops are used to fabricate newly designed equipment/inventions.*

Computers have simplified the process in some aspects, but making a machine that physically performs a function always involves:

1. Arrive at a concept
2. Convey to others what the object will look like

3. Make drawings of all parts to be fabricated
4. Assemble the parts
5. Design the prototype
6. Establish final part and assembly drawing
7. Write a specification for manufacture

Many of these steps can be done on a computer where changes are easier to make.

When I was in engineering school, we had courses in all of these steps. And we had to fabricate parts ourselves in the school machine shop. We even had to make metal castings. In 2020 very few engineering schools still require learning how to make traditional engineering drawings and learning how to run a lathe, mill, shaper, surface grinder, drill press, cylindrical grinder, arc welding and sand casting, but most designs cannot be implemented without fabrication in some fashion. Additive manufacturing (3-d printing or AM)) can make a part without machining, but that process is still very much in the development phase.

Skilled trades – n., a collective term for people who do the execution involved in building facilities and things

The skilled trades that relate to present-day engineering include:

Mason – n., a person who makes things with concrete, bricks, stone and related inorganic materials

Programmer- n., in manufacturing, most machine tools are computer controlled and this person inputs the required steps

Plumber- n., a person who provides the means for fluids to flow to and from structures

Sheet metal person- n., a person with the skills needed to make things from sheet metal products

Rigger- n., in manufacturing, a person with the skills to move and install large machines

Iron worker- n., a person with the skills to assemble beams and the like to create a steel building

Welder, n., a person skilled in using melting of metal to join and fabricate metals

There are many more skilled trades and they vary with the type of industry, but those listed are some of the skilled trades needed for manufacturing. Most of these trades have apprenticeships associated with them. In most of the USA, state education departments oversee apprenticeship programs. A program may last 4 or more years where apprentices are required to log their education and work hours. An employer usually coordinates the program. When an apprenticeship is completed, the apprentice is granted a certificate that allows him or her to seek employment as a "journeyman" tradesperson at another company or he or she can stay where they completed their apprenticeship. Sometimes apprentice programs in the USA are offered by labor unions. Large manufacturing companies usually have apprentices in several of the trades. Apprentices are essential to maintaining a skilled-trades workforce and engineers need the skilled trades to build their designs.

Science- n., acquired knowledge on how and why things happen and the continued search to explain what is yet unknown

Throughout the 2019/2020 pandemic, the government leaders kept stating, "we are following the science on stopping the spread". The "science" is the snippets of understanding obtained by a precious few medical and biological researchers who happened to be studying viral infections at the start of the pandemic. Each researcher had an opinion with data to support what he or she offered up, and the collection of opinions based upon studies became the "Science of Covid-19".

At one time, the "science" of our planet was that the sun rotated about the earth, not the other way around as we now know. Thus, science is not sacrosanct. It is what is generally believed about something at one point in time. In my engineering specialty, tribology, it had been accepted that when a solid wears it does so mostly by abrasion, like using sandpaper on something or by adhesion; one material adhering to a contacting substance in a thin "film". Very recent studies in my lab suggest the adhesive transfer is a more likely mechanism than abrasion, which produces scratches and wear particles. In any case, the existing "science" conflicts with my test results. Thus, science is not a fixed body of knowledge on a subject but a growing body that only grows by the action of a "champion", someone with a new explanation.

Research and development (R&D)- n., the use of fundamental learnings to allow the creation of saleable product

Every large manufacturing company will have an R&D function. Most include engineers. When I worked in the photographic film industry, my employer had a research lab composed mostly of chemists and physicists who performed countless experiments on how to form images from light. In the 1960's they worked on understanding how to make colors brighter and sharper or how to form images faster. A new photosensitive film coating would be the new product. Engineers would then be assigned to find ways to coat the new emulsions on a flexible support to make a saleable film product.

Later, my company's R&D evolved into using electronic means of capturing images. The result was digital photography, the photography that now produces billions of images each day on the part of billions of users. The personal computer was the product of related company's reach into ways of making printed documents, copying. Thus, two of the most important inventions in the history of mankind, digital photography and the PC, came from R&D efforts by organizing scientists and engineers in manufacturing companies. Of course, both involved efforts of a number or people; most inventions are a team effort.

There are often two types of research performed in R&D labs: basic or fundamental and applied. Working on new concepts, like making chemicals emit light is fundamental research. Finding ways to make money on new concepts is applied research. Development is making research concepts into saleable products. I spent most of my career in fundamental research in tribology, investigating the technological properties of engineering materials to find ones that solve limiting problems.

Invention-n., the creation of a useful thing by a person or persons not known to previously exist

As mentioned in our introduction, inventions happened in prehistoric times. For the past two hundred years or so, inventions have to be "approved" by a patent by regulating bodies in each country. I was granted three patents when I worked in the chemical process industry. They are non-consequential, but at the time my employer was eager to patent anything that would "pass muster". There are criteria that apply to getting a US patent.

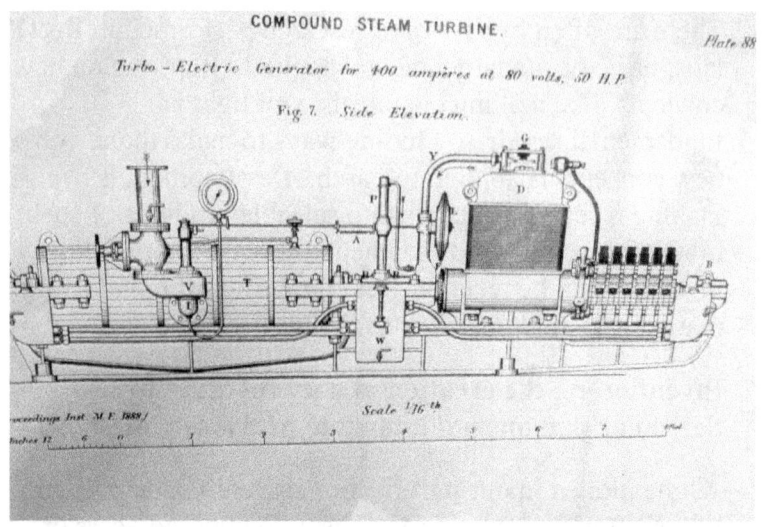

Inside the image:

COMPOUND STEAM TURBINE.

Plate 88

Turbo - Electric Generator for 400 ampères at 80 volts, 50 H.P.

Fig. 7. Side Elevation.

Scale 1/16th

Figure 5 *Patents in the USA assign ownership to a design or process. The government requires submission of detailed drawings and descriptions of what is being claimed in a patent application. This is a typical patent drawing from 1888. Detailed drawings are still required, but today's engineers invariably use computers to make them. Making a drawing of an idea is the first step in the engineering design process.*

Patents are expensive from the paperwork standpoint. When I got one patent I was told that the paperwork and filing costs were about $30,000. However, when you work for a large company, they pay these costs, but they also own the patent even though your name is on it. You may get a token bonus like $500 for each patent. Overall, a patent protects inventors from others stealing their ideas, but most engineers do not patent their design of things unless it is for a saleable product. If an engineer works for a large company the company's position on patenting prevails. Your employer owns the patent of your engineering talent.

There are legal documents that you sign at hiring that deal with patents and other legal issues like company secrets and quitting to go to a competing company. Some engineers do engineering on a consulting basis and thus can own their patents, but over my sixty years in the business I have only dealt with a handful. Most engineers work for a multi-person company or start their own multi-person company.

Conceive- v., to formulate something in one's mind

Where do concepts come from? Of course, they originate in our brains as does everything that we know and understand. Something will trigger a bunch of neuron firings and an idea is born. Product, process or structure concepts usually develop from thinking deeply about a problem at hand.

Build- v., to make an object from components

The planet's early engineers were the people who built bridges over streams with fallen trees, and paved roads with stones, and transported water with ditches and troughs. How did the ancient people on Easter Island raise stones, weighing many tons and meters in size, in the air and set them on other huge stones that they raised up without the help of machinery? How did the Egyptians build flourishing cities in a barren desert? Building is the execution phase of engineering. All of the building wonders of the ancients like the Egyptian pyramids, the Roman aqueducts, the Great Wall of China etc. were conceived, designed, and built by a process that remains the same today. We have more tools to help but the engineering process remains the same.

The engineering process is simple; what is being engineered may not be simple.

Manufacture- v., to make something, usually with the use of others and machines

The manufacture of most of the consumables that we all use like toothpaste, soap, cereal is accomplished usually in factories which are buildings where manufacturing takes place. I have never been to a toothpaste factory, but no doubt that they have machines to mix the ingredients and for sure some automated system for putting the toothpaste in tubes and packaging it. Making soap is similar. The "making" takes place in large vessels and packaging it into individual units is a large effort. Making cereal probably involves some sort of heating process and some way to treat the starting material, some form of grain.

When the USA had a manufacturing economy, maybe twenty percent of the population worked in factories that manufactured the things that we need. Engineers designed the factories, the processing equipment in the factories, the packaging equipment and likely worked on development of new types of toothpaste, soap and cereal. When a thing that we need is made in another country, the engineering needs are met by engineers in that country.

Prototype- n., in engineering, a first attempt at building something to assess the operational success of it

When an engineer designs something, he or she makes sure that it works on paper. In 2020, computer models are common. If everything looks like the "thing" will work as designed, a single or limited number will be built to make sure that everything works. Whenever a new automobile model is designed, they often craft a prototype to see if something was missed in making the design, calculations or drawings. Very often the prototype will uncover problems that would be significant if not addressed.

I believe that my prototypes always showed something that was not right and needed fixing. One time I was involved in developing a new type of plastic molding machine. The prototype was large and complicated. It cost over a million dollars. Also, it did not work. The whole concept fell flat; its parts experienced sticking in one of the forty or so molds in the machine. The entire development was scrapped based upon performance problems with the prototype.

Prototypes are good and necessary.

Model- v., to simulate a process or things operational success with mathematics, and computer assisted images.

Modeling is a significant engineering activity. Most processes show trends. Mathematics can be developed to predict expected results when an event happens. The daily weather report on TV is the product of computer modeling of metrological data to predict a weather result. Engineers develop these models. Finite elementmodeling is a required course in many engineering schools. It is used to analyze forces and stresses in structures and identify weak spots and areas where failures could occur. Some computer models can eliminate costly testing. The lab where I work performs solid-particle impingement test on materials, for example, to simulate erosion on a helicopter blade in use in sand-storm conditions. A research lab in Germany had developed a model to simulate the possible damage modes for every grain of sand. The model can produce the same erosion data as our laboratory (sand blast-type of tests).

Units- n., in engineering, agreed-to quantity designation for measured properties and parameters.

Units are a big deal in engineering. Buildings have collapsed because of errors in converting stresses in psi to stresses in megapascals. I almost failed freshman chemistry at GMI over what is called "unit cancellation". I finally got it and got an A in the course, but engineers have to continually be aware of errors in converting one set of units to another. Most Americans are comfortable in dealing with temperature in Fahrenheit units. However, most of the world uses centigrade. Unit conversion is now available on every computer, but often computer conversions do not have conversions we are seeking out. You have to go back to old-fashioned unit conversion.

In the research community, we have to have all measurements in SI units. "SI" means" Societie International", and this organization sets the standards for use of units in technical literature. However, some of their edicts can be hard to swallow. For example, most people can accept units of rpm for the number of repetitions that a car crankshaft does each minute. The SI unit is reciprocal seconds. Most vehicle drivers would have an issue with a reciprocal-second meter on their dashboard.

In any case, units must be dealt with in engineering and engineers must not make mistakes in going from one system of units to another.

Figure 6 *Architects design buildings; engineers get involved in structural details when designs get complex*

like this office building in Beijing. Licensed professional engineers (PE's) trained in skyscraper design are used to develop the structural details for the design. These engineers along with the architect are responsible for the safely and integrity of the building.

Summary

There are encyclopedias of terms that apply to each type of engineering. We have a chapter dedicated to types of engineering, but the objective of this chapter was to introduce career-seekers to the terms that apply to all types of engineering. We have lots of jargon in engineering, but it certainly is less than the jargon of the medical profession. At least engineering terms usually do not have lots of terms based upon the ancient language of Latin. They are quite manageable and they are not to be feared. None are a disease.

Chapter 3
What do engineers do?

The title question has a simple answer that applies to all types of engineering: Engineers do whatever their employers want them to do.

Of course, that "whatever" is engineering work. I ran into "whatever the boss wants" very early in my career. I have to admit that when I got accepted in cooperative engineering school, I had no idea what I would be doing. My first year's "co-op" assignment in the GM carburetor factory included a stint on the Buick carburetor assembly line, two months as a suggestion investigator, and two months sharpening punch press dies in the tool room. In my second year, I was assigned to the plant's metallurgical lab. I loved it and during my assignment there the chief engineer called me in and asked if I would alter my school work to become an expert in engineering materials to solve a forty percent scrap problem in die casting Corvair carburetor float bowls from aluminum. Of course, I accepted his request and materials engineering became my career.

Many engineering careers start with an early job or assignment that matches what you are good at with what you like to do. The purpose of this chapter is to present the details of a typical engineering career in manufacturing. The details will mostly be from my career and careers of close associates. The chapter's objective is to provide an accurate picture of the everyday activities of an engineer. I started with one of the largest companies in the world at that time (1960's) with over 400,000 employees. Then I worked for one with 120,000 employees when I joined, and now I work for an engineering test lab with two employees, my youngest son and myself. It is his company; I am mostly a supplier of advice, when asked for it. The chapter format is: work during school, your first job, mentor and orientation, design engineering, communication, and career advancement.

Work during school

If you attend a cooperative engineering school like I did, half of your engineering education will be spent working in some capacity in a company who needs engineers. My alma mater, General Motors Institute –GMI, which is now Kettering University, makes the work phase part of the school. At GMI we had to have a significant project during each work block and we wrote a report on what we accomplished. We got a grade from the factory department and from the English department on our "coordination report".

When I started GMI, we were 2500 engineers in mechanical, industrial and electrical engineering. Now there are more engineering options and students no longer work only in GM plants, but in scores of industries. Rochester Institute of Technology (RIT) is another co-op engineering school that has students in many different companies. Of course, I am biased towards co-op engineering programs because they provide a glimpse of your daily work while there is still ample time to move out of something that may not be the right fit for you. Some companies offer summer internships that let engineering students check out working for particular companies.

In my five years as a cooperative engineering student at a GM carburetor factory, I worked in every department of the 3000-person plant. When I worked in the die casting department, I ran a molten zinc die casting machine. The ambient temperature at the machine was about 105 degrees F. When I worked in plant maintenance, I followed a plumber and did plumbing "dirty work". When I worked on assembly lines, I was usually a set-up person. I kept the hoppers filled for the thirty women on the line. They called me "Sonny". When I worked in the metallurgical lab, I did sample polishing and microscopy. So, the work part of the co-op program was mostly real work, but I learned so much about everything. Not only did I learn how things are made, I learned how to work with different people. I learned the intricacies of different factory departments. It was beyond wonderful.

The message we need to convey is that there is no replacement for getting exposure to as many aspects of engineering as possible as early as possible. High school is the place to start finding summer work that relates to your career choice. For engineering, any work that involves building something is appropriate. In the USA, you need to be at least eighteen years old to work in a factory. Thus, for your junior and senior years in high school you may not get factory work, but any related work helps. My high school junior-year summer was spent learning auto body repair. I smashed up my beautiful 1952 Mercury convertible and had to learn body work to fix it. As it turned out, the work was related to my future schooling in metallurgy. I had to learn how to form sheet metal, how to fill dents with molten lead, how to weld and learn how to spray paint. I used these skills throughout life.

A young mind is like a sponge that never becomes saturated. Strive to learn at every opportunity. Learn everything what you can that relates in any way to a career. In 2022, all jobs require computer literacy. If you cannot find a summer job that relates to engineering, teach yourself a computer skill. Maybe make a U-tube video of a unique repair method or go to a library and learn 3-D printing. Most public libraries have this equipment and it has become a significant part of engineering. It is called additive manufacturing (AM), in engineering jargon.

Strive to learn career-related information as early as you can.

Your first job

My first engineering job was my co-op engineering thesis. My fifth year at GMI was full time work and I was assigned a real engineering problem that I had to solve. Mine was solving a forty-percent scrap problem in die casting Corvair carburetor bowls. Because I co-oped at the plant for four years, I did not have to go through an orientation on how the company works and what I would be doing. I had already spent several work blocks in the department where I was assigned the problem. My first year as a newbie engineer did not work out the way that I anticipated. I finished school on August 10. I got married on August 26 and on September 10 the boss called me in the office and told me that starting next Monday I would be working on my thesis-project at the GM Technical Center in Detroit. I would fly down every Monday and back on Friday. I would be using their analog computer to learn how to apply directional solidification to solve our casting solidification shrinkage problem.

How's that for a job surprise? One week after my honeymoon I leave my bride for four days each week. I was on an expense account indefinitely, the company paid my plane fare, my hotel, my meals, my valet service (they gave me a car to use in Detroit), and my laundry. I'll go into this project a bit more later, but since my new wife was working at that time, we treasured every weekend. I also learned that other engineers spent their week away from home.

My Detroit travel lasted most of my fifth year. I had a standing plane reservation for every Monday and every Friday and after a few weeks I noticed the same faces on the plane on Monday morning. The plane that I took originated in Boston; I picked it up in upstate NY and the plane went on to Detroit and then Chicago. I got to know some of the "same faces"; they were sales engineers who had territories that took on the northeast USA. A sales engineer learns the details of a product or process and then is assigned to calling on potential customers to sell the product or spends his or her time dealing with technical issues related to the product or process. Some products are complicated and need an engineer for application solutions. Sales engineering has always been a significant part of engineering.

Mentor/orientation

On your first job out of school or anytime that you change jobs you will likely need a mentor to acquaint you with how your new company works. When my co-op work evolved into a specific field of engineering (die casting), I was assigned to a senior engineer for mentoring. Newbies in any career field are not of much value until they acclimate to the company.

After graduate school, I changed companies. My new job was at the largest manufacturing facility in the world at that time (1964) with over 30,000 employees at a site that covered hundreds of acres and had hundreds of buildings. This company fortunately provided a mentor for the first year. My mentor was the welding engineer in the Materials Engineering Laboratory where I was hired. Al was a company "lifer". He spent over twenty years doing welding engineering and he knew the huge facility like the back of his hand. He knew everybody and I "tagged" along when he went into the field to solve welding problems. He taught me how to navigate the site, the relevant contacts, the nature of his job, and how to get along with the thirty or so engineers and technicians in the lab.

Having a mentor is a huge help in starting a career.

I have no idea how I would deal with such a huge facility without a mentor. And he was one of the nicest persons that I ever encountered. It took me about two years to get comfortable to go on jobs by myself. Most projects during my first two years were assigned by the lab boss. After that I was on my own. Material engineers were consultants to the entire corporation in their specialty, and as such, I spent my remaining thirty-six years with the company establishing my own research and development projects.

Design engineering

Many, maybe most, engineering jobs involve designing something that is then built. I started my engineering career "on the boards". I was given a very large drawing board (maybe 4' by 6') with a drafting machine and was told to design an automatic die casting machine part extraction for a particular carburetor part. At present, the casting is removed from the mold by an operator using a pair of plumbing pliers. I was to design a device to eliminate an operator and that was the cost justification for the design work.

The steps in design engineering are pretty much the same for anything that is to be engineered:

1. Receive the assignment

2. Establish a design concept
3. Make an assembly drawing of the device
4. Get details drawn for each part
5. Establish needed controls/data collection etc.
6. Get details machined
7. Assemble the machine and debug it
8. Modify design as needed
9. Test final design in production
10. Sign-off on design, patent the design if justified

This may sound like a huge task. It could be, but my automatic casting unloader was not something brand new. The plant had maybe fifty zinc die casting machines and six aluminum die casters. There were automatic unloaders working on about six machines. Thus, I made my design concept the same as the automatic unloaders on other parts. All that I really had to do was to figure out how to grab the part involved in my assignment. It was a cluster of eight throttle control covers that resembled lids for canning jars. The eight parts were held together by metal feed paths connected to a central metal injection site called a sprue. I just had to develop a clamp for the sprue.

A molding cycle took about 60 seconds with about 20 seconds of that cycle allowed for removing the part from the mold when the mold opens. The cluster of parts is pushed out of canisters by ejector pins. My design assignment started by obtaining full size drawings of the assembled mold and part cluster. I made a full size drawing with the mold open with the part pushed out on the ejector pins. Then I sketched many mechanisms that might work to clamp the part. Once I found one that "worked on paper" I coupled it to a mechanism similar to what we were using on other die cast machines. When I felt like I had a design that would work I asked my mentor for his opinion. He did not like some things and I changed them. When he thought my design was okay, I sought approval of the design from our boss. Once I had the design approved, the individual components of the design could be "detailed" by drafters.

What is a drafter? When I started on the "boards" a typical engineering department was a large open area with ten to fifty drawing boards. The engineers may have had a drawing board and a desk. The draftsmen (now drafters) were non-degreed people who could draw and print well. They took my assembly drawing and made detailed drawings which means enough views and dimensions that a machinist who was completely unfamiliar with the device/machine could make the part. The engineer was responsible for specifying what the part would be made from, if it needed any treatments like heat treating or plating and all surface finish requirements.

The assembly drawing and detailing is now done on computers and it is also possible with computerized machine tools to skip having a machinist interpret a detail drawing. The detail can be electronically inputted to the machine tool and it will make the part.

Three-dimension computer drawings can now be inputted to AM machines and the part made without cutting it from a block of material.

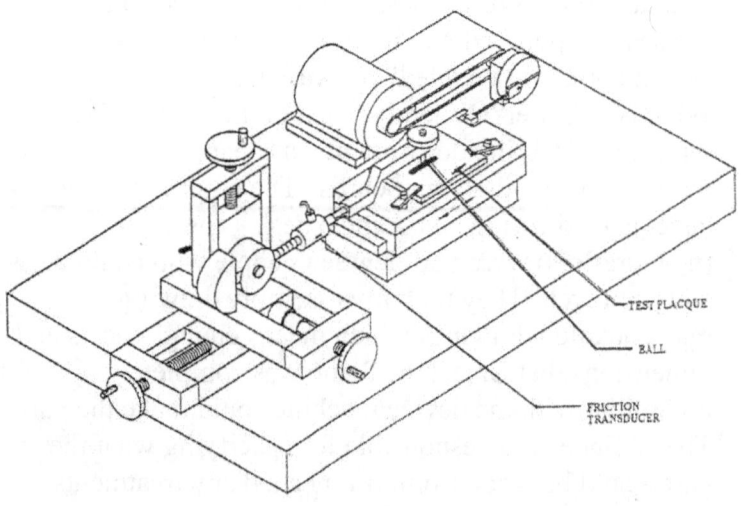

TEST PLACQUE

BALL

FRICTION
TRANSDUCER

Figure 7 *Three dimensional views of a machine concept are generated from two-dimensional computer drawings made by a draftsperson or engineer.*
Drawings like this are often used by engineers to present design proposals to managers for their approval.

After details are made, a machinist will assemble everything and try it. If it looks like it will work, a "debugging" step will follow. In the case of my part unloader, it was soon in production with somebody watching how it worked. The "watcher" was often an industrial engineer. Industrial engineering is a degree option in many engineering schools. My undergraduate degree was in mechanical engineering. That is why I designed the part unloader. An industrial engineer will study the ability of the device to replace an operator and save the money that it was supposed to. Industrial engineers study processes, methods, quality control, and operations.

This can be hugely important. I am writing this on an airplane. After we were loaded onto the plane and ready to go, the auxiliary power that supplies light and heat at the terminal gate failed. It took the ground crew two hours to restore the power and get us underway. People were running in and out of the plane cockpit. The plane doors were opened and the stewardesses were looking out. It was complete chaos. Nobody established a procedure for this kind of failure and it cost 160 people two hours and many missed connecting flights. Industrial engineers establish methods and procedures to keep something functioning and develop ways for making things work better.

The industrial engineer proposed some tweaks to my unloader, we made them and did another debug run. Finally, I got mentor approval and supervisor approval. Then I had to document the whole project: what was done, the results, the savings. This device did not warrant a patent since it was not a unique entity and did not warrant the $30,000 cost or a patent application.

In summary, design engineering usually requires quite a few steps. If the design is for an electrical device or a chemical process the steps will be a lot different in nature, but probably no less in number. Going from concept to device/product/process takes lots of thinking, lots of steps, and lots of problem solving. This is what design engineering is like.

Communication

Communication is a fundamental part of all engineering. When you finish a project, you have to document what you did, but you will probably have to make oral presentations to your peers and supervisors. Most engineering schools promote taking public speaking training and make report writing part of their or other school aspects.

Just about all employers of engineers have formal systems for getting funding for projects. The larger the company, the more likely that reports and presentations will be a job requirement. I have witnessed brilliant engineers passed over because of poor communication skills.

These skills are easy to acquire in your college work. Most engineering courses that involve laboratory work require a written report on your lab findings. Mandatory engineering courses like chemistry and physics invariably require lab reports for grading on the course. Learning to write well in school courses that require reports and papers will hone the communication skills that you will need in an engineering career. Of course, it will be expected by all employers that you be proficient in the computer software products that are widely used by professionals.

A good engineer is a good communicator.

Research and Development

As I mentioned, my fifth-year thesis was on development of a new aluminum casting process to make Corvair carburetor bowl castings that did not leak. My work at the GM Technical Center in Detroit was pure research. The technical center was looking for a place to apply their analog computer. This was a very large room full of electronics and a model of my casting was simulated electronically on two-dimensional conducting medium (full size) and my job was to physically move around coolant lines as electrodes with an assigned electrical potential to obtain thermal gradients that would allow continuous feeding of molten metal as the part solidified; I had to find a way to feed solidification shrinkage which was the cause of the parts scrap problem; shrinkage cavities causes needed vacuum to be lost. Essentially the computer and my waterline electrodes were solving heat transfer equations using voltage proportional to temperature and current proportional to heat flow.

The project was successful. I spent the whole year on it. However, the Corvair went away as did the analog computer and I moved on to research in tribology.

Part of R & D is getting acceptance of your invention. I invented many devices for controlling sails and rigging on my cruising sailboat, but had no successes marketing. Sales and marketing engineers are an integral part of engineering. Sometimes if you develop something, the only way to make it a success is to market it. In my lifetime, I witnessed lots of good inventions wither because of poor marketing. One of my favorite inventions was the plastic automobile, GM's Saturn. Since I live in the US rust belt, the concept of a car that did not rust to me was something very desirable. History shows that GM's plastic car, the Saturn, did not succeed. Maybe, because they did not market it as rust-free. Or maybe it died because they went back to steel for body panels.

Engineers also have to be wary about marketing something that is not ready to market. As I write this there is a court case in progress on a medical product that was well-marketed, but did not work. Thus, marketing and R & D need to be balanced. If you want to aggressively market a development, be certain that it works.

One of the essential aspects of research and development in any field is searching of records on what had been done by others in your subject of interest. All papers published in peer-reviewed journals have to start with a literature search showing that you researched what was previously done by others and show that what you are writing about advances the body of knowledge on the subject.

The worst thing that can happen to you in presenting a paper at a conference is to have a person stand up during the question period and say "Did you know that xxx did this same experiment in 2010?" You cannot miss important work in your research effort.

Effective literature searches are a core part of R & D. For over twenty years, my lab was next to the site's engineering library. The librarian often did my searches and he knew the best search engines. There are also pay-for-searches available. Whatever method is used, literature searching is an integral part of all aspects of engineering. Knowing how to do this is a career requirement. You usually only need to do this at the start of a project, but if you are patenting something the search for other ways of what you want to do is fundamental to getting a patent.

From concept to product

We described the overall process of engineering design, but many details were omitted. To illustrate all that is involved in engineering a product, building, process etc, we will chronicle the development of a test machine that I developed and which my son's company built for sale.

When I was working in the photographic industry, we were often confronted with wear produced by moving movie film on projectors and the moving film, mostly 35mm-wide movie film, would develop "spots" when projected. The spots were produced by the product of the wear particle from film rubbing on projection equipment. It was common practice at that time to add nanometer-sized particles of hard substances to film coatings to make them more "wear resistant". Some of the tiny particles were chemically aluminum oxide, a known abrasive.

To study this phenomenon, I set up a rig in the materials lab to rub the film against candidate materials, identify and sort abrasive films and to screen candidate materials for resistance to erosion/abrasion in guiding film transport in manufacture and in customer equipment like movie projectors.

Years later when I was then working at my son's testing company, I suggested developing an abrasion tester that would work like the lab film abrasion test I used when working in movie film. The abrasion test rig would use commercial abrasive tape (sandpaper) that is used in producing a smooth finish on automobile engine parts. The abrasive tape is available in different grit sizes and widths.

I hand-sketched what the machine would look like and got design reviews from my son and potential users of the machine. When the reviews agreed that the concept might work, I started the design process on a large sheet of drawing paper on a large drawing board. This, of course, would now be replaced by a computer monitor that would not allow a full-size image. The machine would be about three feet high and two feet wide.

Assembly drawings require enough views to allow all of the components of the machine to be seen in at least one view. When the assembly drawings are complete there is another design review to see if the machine works on paper. Computer-generated assembly drawings can be given motion analysis with some computer-drawing programs. In this step you may find that a lever hits a screw that sticks out and you must change the assembly drawing to designate flush screws.

Each part on the assembly drawing is given a "detail" number. The detail is then described on a list on the assembly drawing for example:
(10)   10-32 socket head cap screw

Thus, the detail can be a purchased part or a part that needs to be fabricated:
(11) left lever, material: 6061 T6 aluminum clear anodize coated

The detail description usually shows the material to be used to make the part and any treatments that are necessary, like heat treating or coating.

A drawing must be made for each detail that cannot be purchased. The detail drawing usually needs several views and in large engineering departments "drafters" will make detail drawings and the engineer signs them to approve them.

Next, details are fabricated in a company's machine shop or they are bid to outside machine shops. When all of the machine details are fabricated, they are sent to a shop for a machine to assemble them into the machine.

Concurrent with the machine mechanical components, drawings are made for the electrical controls or hydraulic or pneumatic controls for the machine. The mechanical design engineer may have the department electrical controls engineer make the electrical drawings. In my company, I made the machine drawings and my son did the controls. He also machined all of the details that had to be made.

Once the mechanical components of the machine are assembled and everything seems to work, the electrical (or other energy distribution) controls are completed. Motors, lights, sensors, computers are wired. Some machines need software written. We had a computer consultant design dedicated software for the machine.

After the controls are in place, debugging begins. Debugging is the term commonly used in the USA to mean: turn it on and see if everything words and fix what does not work. If you are making a number of these machines, as the engineer you may need to write up the entire process and develop an inspection procedure for making more of these machines.

Figure 8 *This machine uses a fixed abrasive film (sandpaper) that is in line contact with a test specimen. A designated test sequence abrades a groove in the test specimen and wear volume on the test specimen is calculated. It becomes the measure of that specimen's abrasion resistance.*

Figure 8 shows the final form of our loop abrasion tester. We made about ten of these, four for ourselves and six for other companies. Many machines are designed and developed for special applications where the world market may be less than one hundred. Making thousands can be done using vendors. Making millions of units will require a factory and engineers will be needed to design and build the machines that make up the factory.

Most types of engineers follow similar procedures but the steps may be different. Civil engineers do not make life size drawings, but they make scale drawings. Computer engineers may develop personal computer code as their product. A materials engineer may develop a new plastic as his or her product and drawings are replaced by interactive plastic recipes and lots of property testing. Biomedical engineers may design a knee prosthesis and do debugging in a patient. The details for engineering depend on the type of engineer and the company (employer) situation.

Career Advancement

An average engineer will have definite milestones in his or her career.

1. Get established within your company (first 5 years)
2. Plot a career path (management or not – 5 - 10 years)
3. Settle into one company or shop around (20 years)

4. Download your learnings to others (10 years)
5. Retire (20 years)

What you do in the first ten years out of school can determine your entire working career. I started doing tribology research about three years after graduate school. I am still doing it fifty-five years later. I will present two papers this year.

My chosen career path was to become a technical expert in engineering materials. I chose this as opposed to the most common engineering alternative: management. Of course, every engineering position has a boss. So, wherever you work you may opt to move up in your career by becoming an engineering manager or manager in some non-engineering function. Early on, I was made a "group" leader which meant supervising about six other engineers. I quickly learned that this route was not for me. I spent most of my day on people problems rather than coming up with product or material innovations. I went back to R & D after about two years. However, management always pays more. If you are a boss, you will likely make more money than not being a boss.

Figure 9 *A typical metal fabrication shop in the 1940's. Engineers would design large machines to make various products like movie film and these skilled-trades people would fabricate and install the machines. Bosses wore white shirts and ties in those days.*

I never gave money a thought in my entire career. Engineers make good salaries and you can have a comfortable life just being an ordinary engineer. I never asked for a raise in my life, but I usually got one when they were available.

Working for "money" is usually not part of engineering.

I go into the lab six days a week to work in R & D relating to my son's tribotesting company. I do not take a salary. I do not need more money since I retired from a big company with a good pension.

Two of my college classmates went on to become CEO's of very large corporations. So, engineering can be a career path to the top of the ladder. You simply have to make your path decision early.

Summary – what do engineers do?

It depends on your situation, but traditional design engineering to make something or some process is like I described. Software development and controls may be different in the steps you take, but the concept of starting with an idea and making something that can help others is basic to all branches of engineering.

Engineers make what people need and want.

Figure *10 Engineers design aircraft for passengers, freight, military, and emergency services. Commercial flying barely existed before World War II. Now planes fly everything everywhere and most ordinary people fly to wherever they like.*

Chapter 4 Important Engineering Achievements

Prehistoric times

What we know about people, who lived before living together became popular, comes from discovered artifacts, graffiti and drawings left by our ancestors. One thing is certain about the people who hunted and gathered to sustain life; it was not an easy life. Archeologists have uncovered the bones of large animals like mastodons with sharpened sticks lodged in their bones. These kinds of discoveries suggest that the hunters and gatherers hunted their protein with sharpened sticks. They burned the ends of the sticks then rubbed them on rocks to sharpen them. Thus, a significant engineering achievement was finding a way to make spears sharp enough to penetrate elephant hide. However, an even more important engineering achievement was the invention of fire and the means to start and maintain fire for cooking and warmth. Someplace in antiquity someone engineered fire starting using a stick as a drill rapidly rotated by a stringed bow. Making vessels from animal parts and sunbaked clay was another seminal invention. Water and food could be stored. Needless to say, many inventions from prehistoric times related to clothing and shelter. Animal skins provided clothing and footwear and many different inventions provided shelter. Natural caves were continuously occupied for centuries. They were the most robust houses, but they may have had to negotiate with bears for ownership. Sometime in prehistoric times early engineers invented "boats" for travel and moving things. They undoubtedly started as floating logs, then rafts, then somebody thought about hollowing out a big log with fire and the canoe was invented. Some early engineer noticed that paddling a canoe was easier with a following wind blowing on his or her back. Why not put some woven palm branches up to enhance the wind's propelling effect: the sail was invented.

Figure 11 *The invention of the cell tower resulted in wider cell phone reception.*

Many engineering projects do not involve something new, but improving on existing inventions. The sharp stick to hunt game was greatly improved by the invention of sharpened stone arrowheads that could be attached to the stick and these improved sticks could be thrown to penetrate rather than the necessity of coming into close contact with a beast. Eventually the spear was reduced in size to form an arrow and the bow and arrow further reduced the need for direct-contact in the daily quest for food.

Someone also engineered tools to scratch images on cave walls. Someone engineered dyes and ink to color cave wall images. Together these innovations gave the world a glimpse of daily life in prehistoric times.

Early civilizations

People started to live together when "farming" was invented. Farming allowed people to stay in one spot and not have to follow game as it moved about. Food for existence was grown and stored. Archeological digs in the Middle East suggest civilizations of several thousand people occurred about 10,000 years ago. A city-state in Mesopotamia unearthed by archeologist's shows that engineers invented houses and fortifications. Incredibly, war, as evidenced by archeological digs of many bodies showing spears in them, progressed along with living together. Many engineering firsts were likely the result of war between competing city-states. For example, horses were a source of food like deer until some engineer devised a way to corral some. Then somebody learned that they could be domesticated and an engineer devised equipment to "steer" the horse; then engineers devised plows and chariots to be pulled by horses and we ended up with a civilization that used these animals to do everything from farming, to transportation, to war, for thousands of years.

One of the inventions that amazed me from early civilizations was the "flip-flop", the footwear held to the foot by a thong between toes. One time I was presenting a paper at a conference in Torino, Italy and stumbled upon a museum near the conference. They had an incredible collection of Egyptian treasures. In a display case I saw sandals (commonly referred to as "flipflops") absolutely identical to those that I use, but made from a different material. Theirs' was woven reeds, mine elastomers. The display dated the sandals at 3200 B.C. How's that for a nice engineering design, over 5000 years and they are still making them to his or her design. That is an incredible feat of engineering.

Some engineering feats are transformative. The use of horses as work animals transformed civilization. Most civilizations had some form of spirituality and engineers evolved to build their temples and ritual locations. The Egyptians left us carved images of many people pulling sledges with huge stones for their pyramids. The civilizations with domesticated horses had them do this kind of work. Our milk was delivered by a horse-drawn wagon. Thus, horses and the means to use them for work was transformative and it lasted for more than 5000 years.

Figure 12 *Horses have pulled wheeled vehicles to transport goods and people from about 3000 BC to about 1950. Some parts of the world still use horses for transport of goods and people. The development of steel as we know it in about 1850, allowed wooden wheels and wagons to be replaced by steel vehicles by about 1920. The horse was mostly replaced by the internal combustion engine. The wheel was an incredible invention by early "engineers" and the horse was and is an incredible gift. Internal combustion engines are the product of "modern" engineers.*

Maybe the greatest invention of human history, farming, had countless inventors. Somebody discovered that you could plant seeds to regrow what you were eating. It probably happened with watermelon. Those of us who love watermelon, but not the huge seeds, often would spit them into the shrubs around the house. Then next year watermelon plants showed up in the shrubs. People found that they could grow food and no longer had to search the forest for sustenance. People could stay in one place. I witnessed how it may have envolved when I visited the Mayan village of Chichen Itza on the Yucatan peninsula of Mexico. Chichen Itza was a thriving city-state that disappeared maybe 900 years earlier. On the bus ride from a coastal resort to the city the bus stopped at straw huts along the way. There were government-mandated speed bumps whenever the roadway was near a Mayan home. The Mayans are still living in the jungle like they did for thousands of years. Their huts were surrounded by jungle, but in the jungle they planted seeds for corn, squash and beans in single location; the corn grew tall to support the beans and squash and allowed Mayans to find their "produce" in the jungle undergrowth. I also saw wild pigs tied to trees near huts. They did not have electricity or refrigeration, so a tethered wild pig or other small animal provided fresh protein without refrigeration.

I suspect that early civilizations "farmed" somewhat like this. Engineers continually found ways to increase food production, thus allowing civilizations in the form of city-states to grow larger. Of course, irrigation was a big part of increased food production. Engineers devised clever ways to bring water to the places where people decided to stay in place and grow food. There is archeological evidence that settlements in the Middle East got their water from dug channels and aqueducts as early as 2000 B.C.

Materials Engineering, my field, was the factor that produced what is known as the Bronze and Iron Age. Metals for tools had to be invented. All of the elements that we now know exist are present in the earth's crust but only two of these elements exist in metallic form: gold and copper. Iron is plentiful in the earth's crust but only in compounds and gleaning iron from its ores is not a simple process. Gold was known and used probably more than 10,000 years ago, but it is soft and malleable, besides being scarce. Thus, its use was primarily for decoration or money. Gold coins allegedly were invented by Croesus, the king of an ancient middle-east state in the first millennium B.C. This invention was transformative. Prior to "money", transactions were by bartered goods or other arrangements. Working for money and buying things with that money changed civilizations. Incredibly, gold is still a basis for some currencies and an important economic commodity.

Copper can be found in nature in metallic form. The college where I got an MS in Metallurgical Engineering, Michigan Tech, is located on a peninsula in the United States' biggest lake, Lake Superior. Copper was mined in this area since about 1850. In the early days of mining in this region, copper occurred in ore form as well as in the form of mass copper. The entire peninsula is very rocky and if you see "green" among the rocks, it would be some mass copper. If you fracture the rock, you could pick out the shape of mass copper. It could be as small as grams or it could be ten tons. The school had a 10-ton "nugget "of pure copper on display in a campus courtyard. "Mining" consisted of digging out the local rocks and smashing them with giant hammer mills. The copper was picked out by hand and the residue, the copper ore, was called "stamp sand" and was used for roadways and the like. It was thrown out. I used to hunt for mass copper in the stamp sand piles at a nearby mine.

I picked up stones with the telltale green deposits and we would extract the mass copper, dissolving the rock in hydrochloric acid.

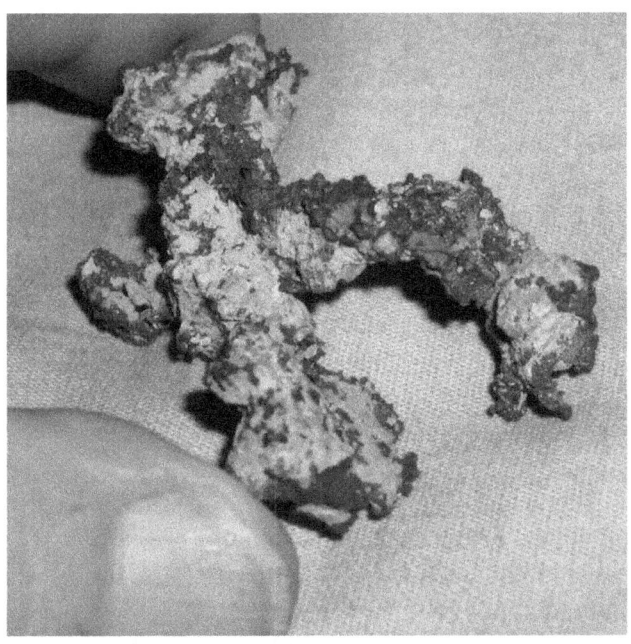

Figure 13 This is a piece of mass copper found in an abandoned mine in northern Michigan in the USA. Copper is one of just a handful of metals that can be extracted from the earth's crust in metallic form. It could be obtained by crushing rocks and picking out pieces of mass copper. This is why it was the first metal that was widely used for tools. Early" metallurgists" learned to mix it with tin to make it harder and more useful. The result was the " bronze age". For centuries "metallurgists/engineers" developed the processes and manufacturing techniques that brought today's world the many steels and other metals that we use to make our amazing machines and devices.

Thus, mass copper provided early civilizations and maybe prehistoric civilizations with the first metal for tools. Unfortunately, pure gold and pure copper are both soft and malleable and not suited to making tools. Some early metallurgist fortunately discovered that when molten copper is mixed with the metal tin, it becomes much harder and copper tin alloys called bronzes were invented. Bronzes made much better tools and weapons and the Bronze Age (3300 to 1200 B.C.) resulted. This was a transformative invention. It quickly spread over much of the world even though tin was scarcer than copper.

Metallurgists, possibly in China, found ways to coax iron from its ore. Some engineer invented the "bellows" which could be used to make hotter fires to allow reduction of iron ore to metallic iron. I saw iron castings in a Shanghai museum dated: 3000 B.C. Thus, at least the Chinese had the ability to cast bronzes and irons into intricate shapes. Metallurgists also discovered wrought iron which contains the slag from the melting process. People started to make tools, weapons and armor from wrought iron by the first millennium B.C. The Iron Age was the result (1200 to 350 BC.)

The only metallic iron on earth before metallurgists learned how to convert ore to iron was meteoric iron. Meteors are few and far between, but early peoples found some and devised ways to cleave off pieces and form them into tools or weapons. Of course, an iron sword could easily do significant damage to a bronze sword. Then other metallurgists invented ways of hardening iron objects and making hardened weapons was a significant industry for thousands of years. Hardening of iron was also a transformative invention.

One of the most transformative inventions of early civilizations was the invention of written languages. The earliest written languages were cut into stone or pottery. There was no paper until Egyptians and others learned how to make inks and write on papyrus sheets and scrolls. The earliest languages were cut into stone, at least that is what survived for archeologists to study. Early people probably started writing on soft stones or clay tablets and used hard stones like flint as writing devices. Some archeologists claim that civilizations without a written language disappeared over time. There was only word of mouth to pass on the learnings of elders.

Greek/Roman/Civilizations

The Greeks invented the Greek alphabet around 800 B.C. that we still use today to name our diseases and many other things. The Romans developed Latin in the first millennium B.C. which is still used as the official language of the Catholic Church. Sanskrit is reportedly the oldest continuously-used language and it was used for the writings of early Indian civilizations. Establishing languages was transformative. Written language allows recording purchases and recording activities of groups of people. Greeks wrote books as early as 500 B.C. This is how we know many things about their civilization. From the engineering standpoint, the Greeks developed a building style that is still popular today. Sometimes it is called "Greek

Revival Style", but the use of large columns in the front of a building with a gabled roof seemingly just sitting on top of the columns was the development of Greek engineers in the first millennium B.C. We use this style on many, if not most, significant government buildings.

Figure 14 Columns with decorations on top and bottom was the classical style used by Greek engineers during the halcyon years of ancient Greece (400-200 BC). Roman engineers later imitated this building style and today countless public and government buildings all over the world still use the "Greek" style for important buildings. Greek engineers got it right.

The Greeks also engineered and built stadiums for public gatherings. Some examples from the great Greek civilizations still stand today. Every Greek city during that era – 400 to 50 B.C. contained temples, public buildings, a stadium/theater and usually some centerpiece building like Athens's Acropolis. They were great engineers and skilled craftsmen. Their engineering achievements allowed them to develop the form of government that we now call democracy, a civilization where citizens participate in running the country. They had to have gathering places for the hundreds of citizens who represented the whole citizenry. They met in "forums" and debated their path forward.

In about 150 B.C. Rome defeated Greece in a big war and Rome spread their civilization to most of the civilizations that existed in what is now Europe and the Middle East. Their engineers built roads to facilitate moving their military as well as goods and services. They also had great engineers who built impressive buildings and infrastructure projects like aqueducts. Of course, they built stadiums in their cities for public events. Many still stand. One time I was stranded in Trieste, Italy by a rail strike and I explored the city a bit and was amazed at the Roman structures, roads and facilities that were still in use. They had a commercial waterfront (on the Mediterranean Sea) that is still in use just the way that it was used by the Romans two thousand years earlier. Roman engineers are often credited with the invention of mortar to allow incredible feats with stone and brick construction. And we still use mortar with bricks in building construction in 2022. The building contributions from the Greek and Roman civilizations are epoch.

The Dark Ages and Medieval Era

After the defeat of Rome's military might around 400 A.D. the civilized world seemed to gravitate into smaller "parochial" civilizations. Countries seemed to not want to conquer their neighbors and rule them at least in what is now Europe. The barbaric tribes that defeated the Roman Empire in the West were constantly warring with one another throughout the early Middle Ages. Some South American civilizations like the Mayans peaked in the complexity of their civilizations in about 1000 A.D. The Chinese and Indians had well-schooled civilizations during what some people term the Dark Ages and medieval times from about 400 A.D. to 1300 A.D. There were not many significant engineering achievements. When there was a need for increased food production horses were made to plow. The horses needed shoes; blacksmiths became a must. They were the first metallurgical engineers. Horses needed leather harnesses and harnesses had to be engineered. Increased grain required mills and ways to convert the grain to flour for baking. Many mills were engineered to use water for power. Once cities started to return, many people lived in clusters. Now engineers had to design and develop fortifications to resist projectiles from cannons. Stone walls at least twenty feet high were no longer adequate. From about the fall of Rome until the European Renaissance the world population was stable at about 100 million. It started growing in about 1600 and has not stopped. It is not known why the world's

population started to significantly increased around 1600
A.D., but education increased and math started to
significantly progress.

Copernicus gave us the order of our solar system. Sir
Isaac Newton gave us gravity. Leonardo DaVinci gave
us the "gear" and countless other inventions.
Universities and learning centers started to develop and
the concept of formal education in universities was
established.

One time I gave a paper at a college in Portugal that
started as an offshoot of several Christian monasteries
which were located near each other. About 1000 A.D.,
the monasteries started a "university" that today is one of
Portugal's biggest and most famous. Centers of
knowledge grew here and there during the more peaceful
times in civilization from about 900 to 1300 A.D.

Besides universities, this time period produced one of
the most significant engineering achievements of all
time: the printing press. Most of what we know about
early civilizations came from books, scrolls and
semblances of books in Greek, Roman, Indian and
Chinese. However, the writers of early "books" had to
get copies written by hand, sometimes by slaves. Later
printing was done with wood blocks of letters wiped
with ink. The invention of the printing press in about
1450 by a German named Gutenberg greatly facilitated
making books. It was transformative, ordinary people
may now borrow or even own printed books.

The Industrial Revolution

Wars seemed to drive technology again from 1500 A.D. on. Wars drove technology before 1500, but what seemed to change things was putting cannons on ships or rather engineering vessels to accept cannon. Many of the great battles/wars from since the invention of cannons in 1300 A.D. or so were naval battles. Civilizations had to develop navies to stave off invaders.

A transformative engineering achievement occurred in about 1700 A.D. with the development of the steam engine. Steam engines began to be used to power all sorts of machines, ships, tractors, harvesting equipment – all sorts of things. This invention may have been a prime mover for the industrial revolution.

In 1793 Eli Whitney invented the cotton gin that greatly speeded up production of cotton for cloth and other uses, but maybe more important he invented the concept of mass production- making machine parts alike for many machines. Parts could be replaced. Each part on a machine could be duplicated instead of each part being unique. This is a big part of engineering: making drawings to allow others to make the same part.

One of the most transformative engineering achievements of all time was the development of the Bessemer process for making steel. It was patented in about 1850 by an English engineer, Charles Bessemer. People had been making tools, armor, cannons and such from iron, but it was not the same as what we now know as steel. Some prehistoric people had iron tools that they made from meteoric iron. Occasionally meteors fall on planet Earth and some contained iron in meteor form. The reentry temperatures smelted the iron from rock. Some ancient peoples found these meteorites and used stone tools to extract the metal. They learned that it could be heated and forged into usable shapes. However, the peoples with meteoric iron were few and far between. Then early peoples found how to smelt iron from ore, but that iron was full of impurities (slag) and could be forged into shapes but it did not have the properties of steel.

The Chinese learned how to make castings from the irons of the day in the B.C. era, but what we now know as cast iron was not invented by an engineer until about 1830. Cast iron contains graphite and sometimes hard phases and its claim to fame is that it is easier to melt than early irons. Cast iron started to be the material of choice for just about everything in the early 1800's: beautiful fences, grills, pots, machine parts even huge bridges were made from cast iron. However, most cast irons are brittle. Iron castings were notorious for breaking under impact and overload. They do not bend; they are not malleable like the wrought irons used for armor and guns. However, from the early 1800's to 1970 or so every city on the planet had foundries that made cast iron shapes for every sort of customer. The "environmentalists" have shut down most of them by the year 2000 (because they had a "smoke stack"), but cast irons will forever be a useful engineering material.

Steels are cast irons with the impurities like carbon/graphite reduced from about 2% to 3% to less than 1%. Getting carbon content down to about 0.2% produces the steels that we all know as the material of construction for the twentieth century cars, bridges, and high-rises. The Bessemer process in about 1850 was the engineering accomplishment that produced this incredible change in lifestyle. Engineers really changed the world with the development of steel.

Figure 15 *Off-road vehicles are needed to build roads, to mine the earth for minerals, and to clear building sites. Engineers design these massive vehicles, their engines, and the facilities to make them. Steels from metallurgical engineers allow these huge vehicles to happen.*

Summary

Throughout history, engineers have developed machines, devices, processes and theories that have changed the world that allowed civilization to exist. All of the things that we use on a daily basis were new inventions at some time in history. Just today I was on a walking tour of new developments in my city and the speaker addressed us standing near a fire hydrant. A person in the group pointed out that the particular fire hydrant that he was standing next to was invented by a local engineer, about 150 years ago and his company made that very hydrant today. This is a perfect example of an incredible invention, one that produced so much good, one that was adapted in most countries and cultures. Cities used to burn to the ground before there were fire hydrants to bring large quantities of water to every building on every street in a city. This example fortifies the point of this chapter:

Engineers throughout history gave us inventions and structures that changed life.

And for the most part, we take things like fire hydrants for granted. This silent, manually-operated mechanism saved countless lives, countless buildings and does so dutifully year after year. Civilization would not be possible without contributions for the good by engineers, past and present.

Figure 16 *This is an electron microscope that produces magnification of images a million times. It is used in all sorts of research from medicine to materials. Optical microscopes invented in the 17th century, which produced magnifications up to 1000 times, allowed an incredible number of research findings.*

Figure 17 *Canals, aqueducts and other water management structures are engineering marvels. They have produced incredible benefits to countless civilizations for thousands of years. For example, the Erie Canal in New York in the USA, was a 350 + mile superhighway for moving goods and people built in 1830 when roads were largely impassible a good portion of the year. Cities grew up along the path of the canal. Now we need more engineers to address water management; too much, and too little water, is becoming common.*

Chapter 5 Are You a "Fit" For an Engineering Career?

What makes you "who you are?"

This is really not a simple question. Every person is a product of genes and life experiences, but there is no formula for the process. I took a DVD course on molecular biology with the hope of understanding DNA and RNA better, but they lost me when they tried to teach me gene sequencing. My takeaway was: they know the chemistry and physics behind inherited characteristic, but they cannot control it without medical modification of genes. If your father had a crook in his nose, it is almost certain that you will also.

One of the personality traits that seem to be established in a person almost at birth is whether a person is prone to be introverted or extroverted. I do not know if this personality attribute is part of transmitted genes, but my personal experience says that it is not. My parents were extroverts, like me, but my sister and brother are introverts. What makes me think that it goes back to birth is that I have been handed many babies during my lifetime and some seem to be at ease with a stranger and some are definitely unhappy being held by a stranger. I think the babies that can be freely passed around at family gatherings are born extroverts and those who do not like to be passed around are introverts.

In any case, being introvert or extrovert can affect one's fit to an engineering career. Some types of engineering careers involve lots of people interaction, some do not. Sales, of course, requires an extrovert, while something like software development leans towards what we used to refer to as "individual contributor" engineering. My specialty has been tribology research and I go to lots of conferences and belong to lots of organizations. My brother, the electrical engineer, spent his 35-year career in the automotive industry designing the motors used for windshield wipers and window lifts in automobiles. Every time a body shape is changed, those motors and their related linkages needed to also be redesigned to fit the new body shape.

Figure 18 *This specially equipped EV was spotted in Banff Alberta Canada in 2023 on its way from the South Pole to the North Pole. Many types of engineers are employed in the design and development of electric vehicles.*

Physical characteristics

Genetics certainly play a strong role in a person's physical size and appearance. If your parents are tall, you will probably be tall, if your parents are large in bone structure, you will probably have a large skeletal frame. Of course, eye, hair and skin coloration are part of gene-related attributes. Most engineering jobs do not have any physical size, strength or coordination requirements, but some do. The ultimate example may be the astronaut. I have not seen the numbers to confirm this, but it has been my observation that a significant fraction of Americans that become astronauts and fly missions are engineers or scientists. They go through grueling physical tests to determine if they can physically take the rigors of weightlessness and confinement in a relatively small vessel. I am claustrophobic so I could never become an astronaut. They also have size restrictions. Being six-foot tall and 250 pounds takes more available space on space vehicles and in the space stations.

In my technical committee work, I met an engineer who worked on oil drilling rigs at sea. He told me that he had to jump into a near-freezing North Sea in a survival suit as part of his training to work on an oil platform. Some people may not have the physical attributes needed for oil platform engineering. I, for one, would not apply for a platform engineering job.

Fortunately, most engineering jobs do not have any physical size, shape or coordination requirements. In my sixty plus years in engineering my most difficult assignment was to climb to the top of a seven-story water tank shaped like a baby food jar and check it for corrosion damage. Climbing the internal steel ladder was okay, but standing on the top with no safety harness was a bit scary. However, one of my nieces is a civil engineer specializing in cell phone towers and she routinely climbs them. I guess that it is an acquired attribute.

Another physical attribute that can affect your fit to an engineering career is tolerance for blood and dealing with internal body organs and the like. My earliest thoughts about a career for myself were about being a doctor. This was when I was maybe seven or eight years old.

However, I later learned that I was squeamish about "blood and guts." I could never kill and dress a chicken, for example, as was the practice at the time with many people in the neighborhood. I got faint at the sight of blood (when I cut myself or saw others bleed from an accident). When it came time for me to go hunting with my companions and their fathers I had to opt out. I could never field dress a deer or, for that matter, shoot one.

There was some part of my brain that prevented me from tolerating dissections, bleeding and things related to the internal functioning of humans. This physical/mental brain limitation prevented me from seeking any career that involved "biological processes". I suspect that this proclivity is inherited since my three sons are like me in this area. One son is worse than me on getting faint at the sight of blood. Neither of us can clean a fish; we do not fish.

Overall, there are not many fields of engineering that have "physical" requirements, but one's tolerance of biological processes can affect a person's fit to certain types of engineering.

Figure 19 *This is a driving simulator that engineers designed to help train race car drivers. They also can be used as driving gaming devices. Computer engineers and electrical engineers are responsible for the electronics of these rigs. Mechanical engineers design the remainder.*

Personality traits

We discussed intro - and extroversion. They are part of the five major personality traits that shape a person. The other four are: emotional stability, openness, conscientiousness and agreeability. Each of these can have some effect on a person's fit with an engineering career.

Emotional stability simply means one's ability to keep calm in crisis and not be changeable in demeanor. Being prone to anger when something did not work as expected would be an example of poor emotional stability. For example, most of my engineering designs never worked "as designed". There was almost always a bolt or nut or location issue that arose at break-in. Engineers are required to accept this and fix the situation and not get angry and tear up the machine drawings. You need emotional stability, patience and calmness when something does not work right at the start.

Openness means the ability to accept new ideas, concepts, ways of doing things, other people and the like. This may be an acquired personality trait. As a young person matures and has new experiences, how does he or she accept and deal with new things. For a young person going from, for example, grammar school to a central high school their surroundings will greatly change. A student becomes exposed to things and situations that they never previously experienced. In my case, I went from two classes of thirty students each in a neighborhood school that I walked to, to an away school of 1200 students. My grammar school had about 500 students and I went through first through eighth grade with the same 60 students. They were my friends, my classmates, my sports-team mates.

When I went to a county-wide high school, rather than a city, neighborhood school, I became confronted with deciding what subjects to take, what clubs and sports to participate in and new ways of doing everything. I tried to be open to many new things. I quickly learned that at 120 pounds I was too puny for the football team and too short for the basketball team and too shy for the debating team. By accident, I learned that I could swim faster than many others and eventually became a varsity swimmer.

Because I went through a cooperative engineering program, half work, half school, each work block introduced me to something new. I had to be open to learning machining, to machine repair, to working on the assembly line, to working in the personnel office. Engineering of all types requires openness. Engineering is a fluid career. Technology continually changes and engineers need to be open to relevant new technologies, methods, equipment – everything.

Conscientiousness – This personality trait usually goes back to the start: infancy. Some infants readily do what they are asked to do by parents, some, not so much. A conscientious person will do an assigned task without supervision and prodding. Two of my three sons would do what I asked them to do without repeat asking; the other needed constant persuasion. When I worked in a large factory job with twenty or so technicians available to do my tests, I quickly identified the ones that were conscientious. They would carry out an assignment to the best of their ability without my constant checking. I could depend that the tests were performed properly and the test data was taken to the test protocol. The job was done right.

It is incumbent upon engineers to be conscientious. Often the consequences of not being conscientious can be significant. For example, a bridge engineer has to check and recheck calculations on every critical member. Lives could be lost by not being conscientious. Most of the time, engineers work unsupervised. You have a supervisor who gives you assignments and rates your work, but he or she likely does not check your design calculations. If something is important enough to need checking by another engineer then conscientiousness requires that it be done. When I started my thirty-eight-year career in the chemical process industry, I knew nothing about corrosion testing. They did not teach this in any of my engineering schools. I had to work under a long-time corrosion engineer before I could trust myself to make recommendations on materials of construction for oxidizing and reducing acids. Using the wrong material gets lots of negative attention. You cannot make a mistake.

Figure 20 *Oops! Something went wrong in this building construction project. Sometimes engineers must play a significant role in the fabrication details of a building that they designed.*

Agreeable

Engineers do not have to be agreeable when it comes to ways of doing things, but they should not be argumentative in all matters. Some people are. Right or wrong, engineers have to be "people" persons. You have to get along with others. For my last twenty years in manufacturing almost every assignment was given to a team rather than to an individual. The team was often intentionally diverse. It would contain, for example, a person from another department. I was often that person. As a materials engineer, I would have the responsibility in the team to ascertain whether the team was using the right materials in a design. I had to get others to agree with my selection, and I had to be agreeable to other teammates' ways of doing things. I had to get along with my teammates.

Agreeableness is sort of carried to the extreme in my work on international testing standards. When I prepare a test standard, I have to ballot it to my immediate committee and resolve all comments and negative notes in the ballot. I have to get people from different companies and countries to agree with the need, usefulness and repeatability of the standard test. Thus, agreeableness is a needed personality trait for engineering. You have to find ways to agree with others and find ways to get people to agree with you.

Beliefs

Beliefs are concepts that may or may not be fact, but you believe them to be factual or true. One of the incentives that I had to be a good student was the belief that a good education would lead to a good-paying job and a good salary and a good salary would give me a good life.

While I was still in grammar school, I thought that money brought happiness. This was my belief at that time. Of course, now I know that that belief was dead wrong, but none the less it was one of my early-in-life beliefs.

Beliefs are shaped by life experiences. One does not need special beliefs to be an engineer, but based upon my lifetime in engineering I have observed that most engineers believe

- that there is always a better way of doing something
- that the best designs are the simplest
- that designs need to be backed up by calculations showing that the design will work when built
- that engineering drawings are necessary and must be checked
- that science trumps opinion
- that engineers have a responsibility to continually improve
- that all people are alike in wants, needs and desires
- that our planet needs caring
- that most problems are solvable
- that people need to share their talents with others

I could probably go on for two or more pages, but the message to a young person seeking a career is that engineering is kind of like a universal religion. I just returned from a conference attended by 1400 engineers from twenty-six countries. We talk the same language and we have the same beliefs in many areas. We are a community. It is really kind of wonderful. We argue concepts, test, techniques and the like, but we all believe that what we are working on will improve something, make something better, or allow something good to happen. This sort of thing happens in many fields, but I think that it is key to engineering. As engineers, we believe that anything is possible. It just needs work and discovery. We can solve the planet's CO^2 problem. We can solve the fossil fuel problem. We just need to work on it and work on it and work together. Thus, beliefs are a part of engineering. We are the external optimists.

Summary – Do your likes match a career in engineering?

Maybe, maybe not. We have presented thoughts on the "mental requirements" for engineering. Candidates for the engineering profession need to take a look at themselves, their psychological compatibility. There are requirements, but they are not onerous and our beliefs are commensurate with living a good life and being a good person.

Chapter 6 – School Requirements for a Career in Engineering

There are school requirements to prepare for a career in engineering and, of course, school requirements to become an engineer. We will discuss both in this chapter.

There is lots of hype in USA educational circles about "STEM": science, technology, engineering and math. Lots of American school systems have STEM programs to encourage young people to pursue careers in these areas. Governments tend to love acronyms for everything and that is likely why "STEM" is being proliferated, but STEM is not a word that I would use by choice. The term "science" can apply to anything. It means a rigorous systematic study of some field. The field can be cooking hamburgers. Technology is a similarly ambiguous word. I once read an essay on how a making corn broom meets all definitions of "high tech". Engineering has an established meaning as we have tried to point out in this book. The "M" in stem meaning "math" is inappropriate in that it may only serve to scare some young people struggling with fractions or algebra. There are precious few careers available in "math". My former secretary had a son who got his PhD in mathematics from a notable university and he spent about two years in "post-doc" temporary university jobs after his doctorate. Usually, college degrees in mathematics only produce careers teaching math to others. With the advent of smartphones and social media, a math major may end up as a "coder", somebody who writes computer code, at a social media company.

What governments and related public education institutions should encourage is careers in engineering and basic sciences (EBS); the basic sciences are: chemistry, physics, biology, and the like. However, the EBS acronym will never replace STEM so young people should accept the government acronym with a grain of salt and accept the E for engineering part of the acronym.

Preschool for engineering

I built workbenches with tools for each of my three sons when they were about three or four years old. They each had their own spot in the basement to build things. Of course, I helped as needed, but mostly I let them teach themselves to make things. I used to take them to exhibits and museums where they could learn more about how things are made and where things come from.

Parents learn very early of the specific aptitudes of each child. My sons all loved their workbenches and their tools. They often used them. There are many educational aids available to introduce preschoolers to math and science. We used to vacation in historic places like colonial settlements where they learned from manned enactments how early settlers lived and how indigenous people lived off the land. These people engineered ways to sustain life from what was available in their surroundings.

Figure 21 This is the lobby of a large hotel with a glass roof. Structural shapes developed through computer engineering models allow building designs that can be new in concept, yet cost effective. The stress on every structural member of the building is calculated and compared to allowable levels with the help of computers. Computers can be an engineer's best friend.

Parents need to accept the responsibility to make preschool years full of things that are fun to do but also educate toddlers in areas that will be helpful in life, and in their careers.

Grammar school

Grammar school is the place where parents should start talking "career" to children. By now parents know the proclivities, abilities, and likes of their children. Now is the time for parents to suggest appropriate careers to their children. However, it is also the time to get a child to read a book like this that explores a potential career.

I visited my father's factory once when I was in grammar school. He was a tool maker in an auto factory. I liked the dynamic atmosphere there and decided I could work in that type of setting.

High School

The purpose of high school should be determination of a lifetime career and preparation for that chosen career path. We had vocation counselors at my high school and determining a suitable life career was a school priority. Career and college selection was pretty much mandated. We had to take many tests for aptitude and compatibility with various careers. In addition, we had talks after school by countless professionals so that we could get details about a wide variety of careers.

Any young person considering a career in engineering or one of the basic sciences (chemistry, physics, and biology) should take all of the chemistry, math, and physics that a school offers. Many high schools offer college-level courses in a variety of areas. My granddaughter took enough of these to allow her to get a bachelor's degree in three years at a prestigious state university. Taking college level courses in high school can significantly affect your college course requirements and definitely should be pursued as long as they do not overload you and bring grades down. You still need good grades to get admission to colleges.

I did not know this until I had been accepted into my cooperative engineering program, but the person that I reported to at my co-op employer told me that one factor that got me into the program was my being captain of my high school swimming team. He told me that he weighs extracurricular activities heavily. Apparently, extracurricular activity participation shows initiative, the ability to learn as part of a team or group, and some level of the extraversion which is required for many types of engineering. Extracurricular activity can be as important as good grades to some college admittance personnel. However, candidates for an engineering career must take all the math, chemistry, and physics offered by your high school. They are all core courses in engineering school.

Figure 22 *This is a bench-top test machine used to study lubricants. Laboratory testers like this are used by most types of engineers to test theories and to compare materials prior to testing them in products. For example, six different manufacturers may offer an oil that is supposed to lower system friction. It may take two years and a lot of money to test six oils in automobile road tests. Screening is done on bench top tests and only the winners in these tests make it to road testing.*

Undergraduate engineering school

I did an internet search of the curriculum for the mechanical engineering program at my undergraduate and graduate schools to see what they are teaching now compared to the courses that I was required to take decades ago. Surprisingly, I found all of the same required courses that I took. The following is a typical four-year curriculum in mechanical engineering.

Year 1 -   Engineering Mechanics
             Statics
             Engineering design tools
             Introduction to programming
             Writing
             Optional humanities
             Chemistry

Year 2 -   Calculus
             Chemistry
             Thermodynamics
             Strength of Materials
             Dynamics of Machines
             Fluid Mechanics

Year 3 -   Electrical Circuits
             Material Science
             System Dynamics
             Natural Science Lab
             Natural Sciences

Year 4 -   Linear Algebra
             Engineering Applications

Heat Transfer
Contemporary Issues

Year 5 -   Senior Design
Applied mathematics
Core electives
Senior Design II

There were some courses in my freshman year that I never heard of, for example, statics and dynamics: I had no idea what these courses were. Statics was about applying forces to various shapes, like load-bearing beams, to calculate the stresses that a material sees in service. Dynamics was kind of a drafting course where you figured out what a lever or cam does as it moves or rotates. It was about analyzing motion in machine members.

Of course, there was math and the basic sciences: physics and chemistry. In 2020 many high schools in the USA offer physics and chemistry, so they may not always be required in an engineering program. I had to take two semesters of chemistry and two semesters of physics. The math requirements in my era started with calculus. At that time, two or three algebras were required as was geometry.

The first year of engineering school sets the stage for your entire engineering career. You soon learn what subjects you like and do well in. In my case, I was indecisive about a major, but at the end of my sophomore year, the chief engineer called me in during my work block and said that they had a significant metallurgical problem and they wanted me to work on it as a priority. They asked me to take as many metallurgy courses as possible and make myself an expert. The school did not have a degree program in metallurgy, but they had a strong metallurgy staff and they set up special classes for me. And that is how I decided on a specific field in engineering. My specialty became engineering materials. I did my fifth year thesis on "Thermal Control of the Die Casting Process" which was about solving a forty percent scrap problem in Corvair carburetor bowls. I ended my fifth year in a cooperative engineering program with a great job and incredibly interesting work.

Figure 23 *Large ships like the Queen Mary, require lots of engineering. Naval architects may design the shape of the ship, but mechanical engineers will develop the propulsion system, the engine, the seals etc.; chemical engineers may design the corrosion-control system; electrical engineers will design the electrical system.*

Graduate School

I had a great engineering job after my fifth year, but I decided to get more schooling in material science. So, I applied at about six schools for an MS in metallurgical engineering. At that time, material science was not a designated degree program in any of the schools that I applied to. Also, at that time (in 1962) most engineering graduate schools had fellowships available for MS and PhD candidates. Fellowships paid tuitions and gave the student a stipend to live on.

I do not remember the specific schools I applied to, but I was married at that time with a baby on the way, so a primary school selection criterion was the availability of married student housing. We chose Michigan Tech because it was the preeminent school in the USA in metallurgy and they had brand new on-campus married student townhouses overlooking a lake and the school ski slope.

Thus, sometimes selecting a school involves more than reviewing their course requirements. I have visited many major universities all over the world. Many had beautiful campuses, but it is best to access the whole package – the courses offered, the financial aspects, the housing, the campus, the city, the travel requirements, the teaching staff, the packages offered by the school.

I was offered no tuition, $3000 per year for my thesis work, and essentially a job doing research for a steel organization who paid my research stipend. I could stay for a PhD if I chose to.

The MS in metallurgy required sixty course credits; the PhD required ninety, but also five years of commitment. The MS could be obtained in two years or less. I had to take at least twelve credit hours per semester. The required courses were mostly metallurgy courses, but I also had to take some math courses and I had to audit undergraduate metallurgy courses that I did not take in my BS program. To stay in the graduate program, I had to maintain a B average. I suspect that this is still the case in most graduate schools.

The specific courses that are required for a graduate degree are different for most schools. Engineering schools that offer MS and PhD degrees have core areas of research that reflect their outside funding. If you want financial support in your graduate studies you will have to do a thesis in the school's area of interests. In my case, my thesis dealt with fundamental strengthening mechanisms in ferrous material. My thesis was "The Effect of Solute Concentration on the Yield Strength of Alpha Iron". It was a very challenging project. Very hard, but I loved it.

Choosing a School

Location – Some young people have never lived away from home; some have been away to camp or stayed with relatives. Being the latter is helpful in determining how you will adjust to being alone and away at eighteen years old. I spent years in the Boy Scouts and I had been away to camps hundreds of miles from home, so I was not fearful of selecting to go to a school that was distant. However, when the day came when I drove to my undergraduate school 350 miles away, I was a bit fearful when I walked into a fraternity house not knowing a soul in a strange city. My fears were short lived and disappeared when I encountered others in my same situation. We bonded to survive in our unfamiliar surroundings.

Some young people are fortunate to have schools nearby that can offer an engineering program that meets your selection criteria. I could have gone local to a similar cooperative engineering program, but I thought that it would be better to follow my brother's path and go to his school which was away. One reason to favor an "away" school is that you will be removed from the distractions that you have in place at home. Engineering school is tough; it will require all of your attention.

Overall, a school's location should not be a significant selection factor. Engineers are alike wherever they are; engineering students are alike in interests and personalities and tend to bond wherever on earth that they go to learn.

Accreditation

Some types of engineering require licensing. You usually need to go to a school that is accredited by the licensing organization. In the USA, engineers who work on public projects like public buildings, infrastructure and the like may need to obtain a professional engineer's license, a "PE license". Each state will have a list of engineering schools that meet that state's accreditation criteria. If you think that you may need a PE license for what you want to do then you need to check a school's accreditation of their engineering program. Most state schools that offer engineering degrees are accredited. The accreditation comes from the American Accreditation Program in the USA.

Living conditions

I already mentioned that I choose my graduate school for their good married student housing, but living conditions in any location can be a school selection factor. It is probably nice to go to schools in warm climates, but I never did that, so I do not know what that would be like. My undergraduate school was in an industrial city with no natural attractions like forests, mountains, lakes, etc. However, I found it to be just fine from the livability standpoint. I only had an hour or two a week for non-school activities so what the factory city in the US "rust belt" offered was fine. It had a small lake for sailing, and German beer and food in a nearby town.

My graduate school was in the harshest weather conditions that are possible in the USA. The school was 500 miles north of Chicago on a peninsula jutting into Lake Superior which never gets over 43 degrees F. We would get snow on Halloween and it stayed until mid-May. We would get maybe 300 inches per year and occasionally the outdoor temperature would reach 40 degrees F below zero. However, the scenery was beautiful; the skiing was great and the occasional 40 below was tolerable. My wife was not an outdoors person, but she was fine because our school housing was snug and heat was included in our $92/month rent. It was very livable.

Like location, the livability of a college town should not be a significant selection factor. Humans have survival genes that teach them to adapt to their environment. You can make any school location livable.

There are many organizations that rate engineering programs in the USA, but they do not serve as licensing accreditation measures. In the USA, the state professional engineering (PE) society can supply a list of accredited engineering programs.

In summary, the schooling requirements for a career in engineering are not onerous. They may look so at first glance because some courses are likely new to you. What is "Statics"? What is Heat Transfer? What is "Finite Element"? Some engineering courses are unlike any that you may have encountered in high school, but do not be disturbed by foreboding course names. I once had to take an elective math course. I chose "Vector and Tensor Analysis", having no idea what a vector or tensor was. As it turned out it had to do with force manipulation on things like rockets. The math regimens were not difficult. Some courses that may be required by a particular school may sound onerous, but most are not.

In my day many engineering schools would have a "zinger", a course with a notorious failure rate. These will be identified by surviving students and they will offer guidance in how to deal with zinger courses. In my undergraduate school the zinger was "Economics". In graduate school it was"P-Chem" (Physical Chemistry). Student survivors' guidance got me through them. Tough courses are the same as tough bosses that will always be part of life. You simply deal with them. However, A dean of engineering at a very prestigious engineering school reviewed this chapter and he told me that engineering schools have eliminated "zinger" courses. Student retention is a priority and if they identify a course with abnormally high failure percentages, they investigate and fix the abnormal failure rate situation.

Schools should be compared in your chosen field first. Then investigate course requirements, schedules, costs, accreditation, living arrangements, etc. Make amenities a last consideration. You can have fun and a good life anyplace. Just choose a school that fits "you".

Figure 24 *Wind turbines for electricity are a priority with many governments wanting clean energy. Wind turbines are a huge engineering challenge. They have an unpredictable driver (the wind), and the operating environment is hostile (sun, wind, rain, salt).*

Chapter 7 Types of Engineering

Choices

It is part of our genetic makeup to prefer some things over others. We all have food preferences, preferences in friends, preferences in activities – preferences in most things. Psychologically our preferences stem from what we do well and what best satisfies our mind, our cognition. People prefer to do what they do well in; and what they like to do. Most professions have specialties. Every job, every activity has good and not-so-good aspects to each individual. A mason may prefer to work with stone rather than bricks. He or she becomes a stone mason. A baker may prefer making fancy pastries over baking bread. He or she becomes a pastry chef. Engineering is also like this and colleges and engineering departments in universities have evolved into different specialties. Most education institutions worldwide have the big five types of Engineering:

Electrical

Mechanical

Computer Engineering

Civil

Chemical

Some schools include:

Industrial

Aerospace

Sustainability

Biomedical

Tribology

Mining

Petroleum

Metallurgy

Materials Science

Needless to say, schools that offer more specialized types of engineering need faculty with backgrounds in these areas and that is why most schools offer the big five because it is easier to obtain faculty in the major fields.

The purpose of this chapter is to present a "sketch" of each type of engineering: what that type of engineer studies, who hires them and what they do. When available, a testimonial from a practicing engineer in each field will be presented. We asked acquaintances from the various types of engineering why they chose that particular type of engineering. Hopefully, replies from successful people may help a young person select an engineering specialty. Sometimes, it can be something other than preference of that field. For example, in today's Wall Street Journal, they interviewed the CEO of a large corporation and asked her why she selected electrical engineering as her choice for her BS degree. Her reply was: "I chose electrical engineering because I heard that it was the hardest engineering curriculum at MIT". In any case, comments from successful engineers will be included wherever we could get one to publish. The chapter objective is to give the reader as much information as possible to help in an early selection of the type of engineering.

The chapter starts with the big five and ends with the less popular (by enrollment numbers) types of engineering. One type of engineering is not more important or significant than the others, but the relative number of students in each field probably correlates with demand. More companies are seeking electrical engineering graduates than companies are seeking petroleum engineers. There are simply less places to work in petroleum engineering. However, for many years petroleum engineers had the highest starting salary. This is how the market for engineers is.

Appendix 1 contains curriculum details for each type of engineering. For example, what courses will I have to take if I elect to pursue a BS in electrical engineering? Of course, the curriculum will vary school-to-school. The appendix shows the course requirements from just one school, but most engineering schools teach the same subjects for a particular type of engineering; they may call them different names and schedule them differently within the four years, but the course requirements in

Appendix 1 are at least illustrative of courses needed for each type of engineering degree. All colleges will also have this type of information available on-line.

Electrical Engineering

When I started my engineering career, we mechanical engineers designed machines to mass produce products. We would start with a clean sheet of paper on a drawing board, maybe draw in the part and what was to be done to it and think about how we would pick it up, move it, drill it etc. Whenever we drew a device to do an operation, we would need a source of motion; a motor, an actuator, rotator etc. In my department, we had several electrical engineers that would design the controls for the mechanical devices and actuators that were needed to make the machine function.

Figure 25 *Autonomous vehicles require lots of electronics, and lots of computer help. This developmental autonomous vehicle has its backseat full of electronics. When autonomous vehicles operate on public roads, all of the necessary electronics and controls need to be downsized and hidden. This is another engineering challenge.*

Electrical engineers would make circuit drawings to go with the machine drawings. In my early years, machines were hard-wired to the electrical engineer's drawings and each factory machine would have a big steel panel box containing all of the major electrical controls. Our engineers had the responsibility to also troubleshoot machine problems in production. If something was not working right on the machine, the electrical engineer was called in to find and solve the problem.

If an electrical engineer was designing an electrically controlled product, like a toaster, there would still be circuit drawings and the same type of troubleshooting to make the device work to expectations. Electrical engineers also design circuits for buildings, even residential houses. In 2022, with the increase in electric vehicles, electrical engineers will be needed to supply power to charging stations. They will also be needed to design the controls in the vehicles.

What is different in 2020 compared to my time in college is that much of the control work will be done with computers, and programming may be an electrical engineer's job.

College courses in electrical engineering had subjects like: circuits, motors, control devices, power systems, microelectronics, control etc. In 2020, electrical engineers may take courses in computer logic, information storage, electronic devices, RF communication, chip manufacture etc; see Appendix 1 for course details. Whatever is in demand in 2020 electronic devices, electrical engineers must learn how to design circuits, analyze them, build circuits and controls and in general develop drawings and instructions to use electricity to power and control needed devices and electrical facilities.

By 2030 most countries will mandate the use of clean energy for the things that we need. Electrical engineers will be called upon to make the conversion from using fossil fuels for most things to using clean electricity for those things.

My brother spent his entire thirty-five-year career in electrical engineering designing the motors and controls used for accessories such as windshield wipers, window and seat lifts etc. He still had to make circuit drawings and motor designs.

Some electrical engineers devote their careers to the design of the electronic devices that now control just about everything. I have had doctors that started out as electrical engineers. Thus, like all professions the basics of engineering are a great start for any career. You are taught to analyze, to create, to solve problems, to document how something should work.

Why I became an Electrical Engineer as my career:

My degree is actually Mechanical Engineer, but I studied all the elements of Electrical Engineering, Circuits, Machines, Controls, Electronics, Transistor Circuits, Solid state Physics and just in case, Nuclear Physics.

I worked in Product Engineering at Delco Products Division of General Motors Corporation producing electric motors for the automotive industry, water pumps for agriculture and rural homes in addition to small AC motors for home and ship use. Developing new products and upgrading old ones, was my job.

We turned electric energy into mechanical work. Windshield wipers, window lifts, power seats, door locks, truck latches, emission controls and fuel saving devices like idle speed control actuators.

Electricity powers everything. Today, everyone carries something running on electricity. That is where the Engineer is needed. Produce it, control it, distribute it, store it, and access it.

Currently, production is not capable of sustaining peak loads in many areas. Environmental issues make increasing production and distribution difficult. Storing is particularly difficult and expensive.

The infrastructure needed to power an electric world does not exist. Building it requires electrical engineers. Making is available 24/7 is essential. Making it affordable is essential. Inquiring, creative minds are needed to make it happen.

Man went into space using mechanical calculators. This is the position the "All Electric World" is in.

John Budinski
Application Engineer Delco Chassis Division GMC.

Mechanical Engineering

I had no idea what mechanical engineering was when I signed up for a cooperative engineering school. My father knew what it involved and he also knew my skills and interests. I soon learned what it was all about when I started working in various factory departments as part of my school work block. As I mentioned earlier, I co-oped out of an automotive carburetor factory of 3000 employees. The company designed and built carburetors for six different automobiles. Some carburetor lines required hundreds of thousands of units per year and each carburetor had at least one hundred parts that all had to be designed and made and assembled. Mechanical engineers designed the carburetors, the machines that made the components, the machines that assembled the components and the equipment needed to test the carburetors to make sure that they would work right in engines.

Mechanical engineers design mechanisms; they design products; they develop manufacturing processes; they design tooling for machines, they design HVAC systems, they run manufacturing facilities; they design vehicles, aircraft and aerospace equipment; they develop military equipment; they analyze structures and buildings for strength requirements; they develop new products, processes, structures and anything else that people need. They are responsible for making just about all hardware and durable goods.

Some of the courses that are in almost all mechanical engineering curricula are: strength of materials, materials engineering, design methodology, heat transfer, fluid mechanics, machine controls, finite element analysis. They take courses that teach how to analyze problems and systems. Lots of colleges assign design projects to junior and senior level students to acclimate them to the design process.

Some schools have only one type of "engineering". It would probably fit into "mechanical" at most other schools. It is a very fundamental engineering choice that can serve as the basis for a specialty. In my case, I moved on to materials engineering, but still use the design skills that I was taught in mechanical engineering. The mechanical engineering choice is probably analogous to "general practitioner" in the medical field.

The kinds of companies that hire mechanical engineers range from drug companies (design production machines) to aerospace companies (design rockets, launch facilities, hardware etc.). Most utility companies employ mechanical engineers for design of transmission facilities. Architectural firms use mechanical engineers to design HVAC systems. Disney World employs hundreds of mechanical engineers to design the workings in their animated characters and features. Any company that needs to design devices that physically handle or transmit forces and motions needs mechanical engineers.

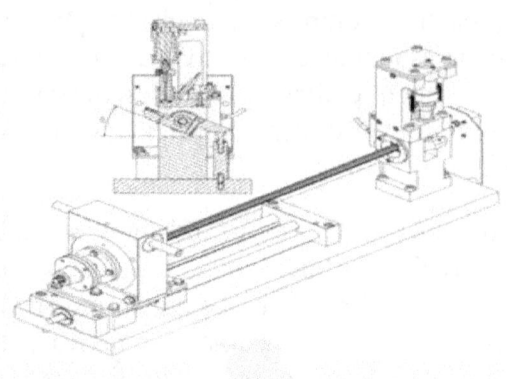

Figure 26 Isometric Image of Phoenix Impact Tester. Making drawings like this are part of the design process.

Why I chose Mechanical Engineering as my career:

My father and grandfather were both mechanical engineers. At school, I was good at physics and mathematics, preferring applied to pure, so an engineering course was an obvious choice. I was lucky enough to study Mechanical Engineering at Cambridge. In a typical eccentric English way, the first two years included courses on structures, materials, electrical engineering, electronics and computing, so a bit more than just the traditional mechanical engineering subjects.

My first job was as a soldier. At age 27, I was in charge of 120 technicians and tradesmen, responsible for keeping 66 tanks on the road, guns front. Tanks might be mechanical beasts, but they also have lots of electrical and electronic systems.

After the army, I joined a start-up called Cameron-Plint Tribology and was introduced to the world of friction and wear. Inevitably, after a few years in engineering design, I got drawn into management. Although financially rewarding, I found management profoundly boring; at the age of fifty, I escaped back into the world of engineering and tribology.

Now, in my 70th year, instead of becoming a superannuated manager, I still have a full-time job in mechanical engineering and tribology, doing things that still interest and excite me!

George Plint MA, CEng, FIMechE: Managing Director, Phoenix-Tribology Ltd.

Royal Military Academy Sandhurst 1973
University of Cambridge 1978

Computer Engineering

This type of engineering has become more popular since the decline of manufacturing in many Western countries. Many service industry jobs are being converted to computers and robots. For example, with the help of artificial intelligence, order taking at fast-food drive-throughs may be taken over by a computer. Computers run the pumps at most gas stations and many restaurants have an electronic device on the table to order and pay for food. These devices will eventually replace waiters and waitresses taking orders.

Computer engineers develop these types of systems. Many schools offer degrees in computer science, computer engineering, and in use of computers for business. Computer science is more theoretical and computer engineering is more hands-on (see courses in the appendix). A degree in the use of computers for dealing with data may be offered by a business department rather than the engineering department at some colleges.

There are many facets to computer science and computer engineering, but all types of computer degrees include learning about software, hardware, networking, programming, and writing code, systems, and applications. Hardware design includes circuit design and other aspects of electrical engineering. Software design includes writing the code or commands that make computers work. For example, one time I was witnessing a demonstration of a person-like robot designed by students at the University of Texas at Austin. The robot was commanded to pick up the candy bar lying on the floor in front of the robot. A computer engineer in the audience asked the person leading the demonstration how many lines of code were required to make the robot pick up the candy bar? She replied: 3100 lines of code. There are lots of details involved in making electronic devices do the work of people. Computer engineers and scientists develop the equipment and the details needed to make computers do useful functions.

Computer engineers work for big and small companies. Sometimes they work on-site, sometimes at home. Of course, much of what they do is done on a computer. Our company started out building test machines for sale. My son and I did the design and fabrication of the machines; our computer engineer did all of the control specifications and software to make the machines run and save test data. This is a typical computer engineering assignment. This type of engineering increases in importance as our civilization gets more complex. If you like to work with computers, computer engineering is a great career candidate.

Figure 27 *Apple IIC computer that cost $4300 in 1983.*

Why I chose Computer Science as my career:

I took the very first computer class offered in my high school in the early 1980s on an Apple II computer learning BASIC programming. I was fascinated with "programming" in BASIC and wanted to learn more.

I majored in Computer Science in the College of Engineering at Texas A&M University with a minor in Marketing. I loved learning multiple computer languages and particularly enjoyed assembly language programming and understanding microprocessor architectures. I loved how coding was both a science and an art in developing various algorithms to solve different problems.

I was lucky to have completed two summer internships at General Electrical Aircraft Engine Group where I was given some very challenging projects including reverse engineering machine code to re-create a lost source code file on a Perkin-Elmer mainframe and participating in a multi-department software development project.

Although I enjoyed programming, I wanted to work more on the business side using both my computer science degree and my marketing minor. I was lucky to have landed a job with Texas Instruments joining an elite training program in semiconductor technical sales after graduating from college.

Working in sales allowed me to work with engineers from various industries including automotive, military, consumer, and industrial electronics. The position in technical sales was exciting to see products go from concept to development, to production.

I ended my career managing U.S. Technical Sales Associate Program for Texas Instruments developing young engineers to become future leaders in sales.

Jane Mulkern
B.S. Computer Science 1988
College of Engineering
Texas A&M University
Strategic Development Manager, Retired
Texas Instruments

Civil Engineering

Civil engineers design buildings, bridges, highways, dams, flood controls and all sorts of infrastructure. Architects are different from civil engineers in that they are more concerned with the aesthetics of a building than in the details of, for example, heating and cooling the building. When I built my custom house, I had architects make the detailed drawings that were given to the framers. Civil engineers were not involved, but the town's civil engineers supervised and reviewed the connections to public utilities, like public water and sewers. New highways and alterations are designed by civil engineers. Civil engineers take some of the same courses as mechanical engineers, like strength of materials and materials engineering, but they also take courses like urban design, bridge design, roadway design, and public safety requirements.

All public transportation organizations hire civil engineers to do road and similar design. Civil engineers who design bridges and public buildings will probably need to have a professional engineer license in the state where they practice. My niece is a civil engineer and her specialty is the design and building of cell towers. These towers are built in many places and each requires adaption to its proposed location.

The bridge and roadway repair that permeates the United States in warm weather months is the product of civil engineers. Governments have civil engineers on staff for infrastructure projects, but much public building and infrastructure work is "farmed out" to private engineering companies. Every city in the USA has a variety of local and larger engineering firms that do public work. More civil engineers work for these firms than government organizations. Mining and energy companies also hire civil engineers.

Figure 28 *Bridges are essential to land transportation. Civil engineers design today's bridges. However, Roman engineers developed bridges that allowed civilization to spread across what is now Europe. Many Roman bridges are still in use after 2000 years. Bridge-building is essential to civilization.*

Why I became a Civil Engineer as my career:

Did you know that bridges move? At learning this, of course, I wanted to learn how this is possible.

For the first year of "take your daughters to work" in 1993, my high school arranged a field trip to a local engineering firm where young engineers presented their projects and fields. I learned civil engineering is the path to learn about bridge design and our building environment. That is the day I decided to be a civil engineer.

Civil engineering teaches about the world around us and allows me to visibly understand how my work benefits society. My career path started traditionally by working as a civil engineer at various architectural-engineering design firms. I have climbed towers to collect data for structural analysis and to verify the tower is safe, walked through desserts and fields to find the perfect sites to construct and traveled throughout the United States to support my projects.

As civil engineers gain experience, licensure and additional education, more paths open to us. I transitioned to an owner representative helping federal clients acquire and receive new buildings, roads, and communication towers to best support their missions.

Today, I am the Practice Area Lead for Capital Infrastructure and Real Estate at LMI, where we support our federal government client's planning, construction and maintenance of the built-environment.

133

Carole Mahady P.E. PMP
State University of New York at Buffalo
B.S. Civil Engineering 1999
Iona College, Hagan School of Business
M.B.A. 2005

Chemical Engineering

Chemical engineering is different from "Chemistry". Chemists are not engineers. Analytical chemistry has to do with identification and analysis of chemicals. Chemical engineering is the field of engineering that designs facilities for the manufacture of chemicals, drugs, water treatment, etc. They design facilities to make sulfuric acid, acetic acid, sodium hydroxide etc. These are chemicals used to make products. Plastic production facilities are designed by chemical engineers. They take courses like process design, process control, physical chemistry; economics etc. (see Appendix1).

I once worked on a chemical engineering team to develop a lithium battery manufacturing facility. Batteries usually involve inorganic chemistry (not carbon related) as opposed to organic chemistry (carbon-based molecules). Any company that makes chemical products (like foods) employ chemical engineers. Water sewage and recycling facilities require chemical engineers. Fertilizer, food and animal feed companies hire chemical engineers as do waste treatment and recovery companies. Chemical engineers are also required by companies that manufacture lubricants and lubricant additions.

Figure 29 Chemical engineers developed the chemicals and spray system to deice airplanes. The sophisticated spray system was developed by mechanical and electrical engineers. The plane and its controls were developed by aerospace engineers.

Industrial Engineering

Industrial engineering is a big part of manufacturing. It is that branch of engineering that deals with making things and doing things in the most efficient manner. In the factory era in the USA, every plant had industrial engineers who would study every phase of an operation or manufacturing process. As a student, I spent a work block in an industrial engineering group. One of my first assignments was to study a machine that made throttle-valve assemblies for single jet carburetors. The machine that made these assemblies had six stations for machining and for inserting parts; the machine operator loaded parts manually. My boss gave me a clipboard and told me to spend a forty-hour work week observing the machine and operator in the day shift. I wrote down start and stop times and noted all operator actions. The machine was not meeting design production outputs. At the end of the week, I tallied machine downtimes and the reasons. The problem became evident from my observations. The operator was spending about three hours a day freeing up stuck parts on one of the machine stations. The machine designers were alerted; they redesigned the parts feeding devices and the machine achieved its design production rate. This is how industrial engineering works.

Students of industrial engineering take courses in time and motion studies, manufacturing processes, statistics, reliability, economics, and data analysis. Every process of any sort could probably benefit from an industrial engineer's review. For example, I stop for coffee and a donut on my way to work every day. The person taking my order has to punch a computer screen twenty-two times to record "one black coffee and one old-fashioned donut". Surely this is not efficient. The food handling is also not efficient. This fast-food facility needs an efficiency study.

Industrial engineering is the field of engineering for people who like things done in the right and in an efficient and cost effective way.

Figure 30 *This is a glass-blowing factory in 1996. There are twelve people within a few feet of each other, and each is handling red-hot glass. An industrial engineer could study this operation and redesign it for fewer people and greater safety.*

Materials Engineering/Metallurgical Engineering

Materials Engineering did not exist as an option in engineering school when I started my schooling in the 1960's. However, most engineering students had to take a course in metallurgy. Metals were the most important engineering material. We had plastics and ceramics at that time, but they were not widely used in machine and structure design. I was asked by my co-op employer to specialize in metallurgical engineering. So, I took all of the courses offered in that area. We had courses in foundry practice, in cast irons, in machining processes, in strength of materials, heat transfer, and the like to teach about materials. My MS was in metallurgical engineering and I went to a school that was the leading school in this field in the USA at that time based on number of undergraduates in "metallurgical engineering".

I was hired after graduation as a metallurgical engineer in what was the largest manufacturing plant in the world at that time (1964) with about 35,000 employees at a single site. I soon learned that I needed to learn more about plastics and ceramics since metals are not always the best material of construction in the process industry. I took evening graduate school courses in plastics and corrosion and after about five years of self-education, I considered myself a materials engineer. I eventually taught materials engineering as adjunct faculty at two local colleges. I wrote a textbook on "Engineering

Materials: Properties and Selection" in 1979 and after nine editions it is still being used in various countries.

In 2022, many engineering schools offer degrees in materials engineering. Course requirements (See Appendix 1) include polymers, composites, metals, ceramics, even nanomaterials. Every company that manufactures engineering materials will employ materials engineers in product (metal alloys, castings, plastics, coatings, treatments etc.) development. Companies that make machines, vehicles and equipment of any type will need materials engineers to identify and test appropriate materials for an application. I spent thirty years recommending materials of construction and solving material application problems. The largest industry in 2022 in the USA, health care, is rife with materials engineers, working on materials for medical devices, materials for medical products, and materials for machines like blood analyzers that are used in health care. Electronics industries need materials engineers for materials of construction of devices. Materials engineering becomes more important each year as we deplete the planet's supply of essential elements. Copper will not last indefinitely.

Figure 31 *This is what is under the hood in a fuel- cell powered automobile. There are lots of electrical devices involved developed by electrical engineers.*

Why I chose Material Science as my career:

Originally, I planned to become an orthodontist: I had good manual dexterity and liked science. I studied mechanical engineering in undergraduate school and thinking that if I don't get into dental school, I can become a mechanical engineer. I found I really liked mechanical engineering, particularly the calculation of stress in components and ended up staying in engineering.

As a senior in mechanical engineering, I realized I knew all aspects of designing mechanical components and structures, but I did not know what to make them out of. So I earned a master degree in metallurgical engineering with a focus on material failure modes: corrosion, fracture and wear. This approach has served me well, I have managed large research groups for automotive poweretrains and have 44 patents worth millions to the companies I have worked at.

At present, I lead an organization in the federal government that determines the causes of failure of mechanical components in national-level transportation accidents such as bridge collapses, train derailments, pipeline explosions, and aviation crashes – a field sometimes known as forensic metallurgy.

With an international reputation in material failure analysis, I publish regularly in journals and give lectures at academic conferences to bring up the next generation of engineers.

Michael Budinski

B.S. Mechanical Engineering, State University of Buffalo
M.S. Metallurgical Engineering Ohio State University
Lab Head U.S. National Transportation Safety Board

Mining Engineering

My graduate school, Michigan Technological University, started in 1883 as Michigan College of Mines. The industrialization of the planet required mining of the metals and other substances needed for products, machines and structures. All of the metals that are now critical to mankind's existence are mined. The most important material of construction of infrastructure, concrete, has components that are mined. The lithium needed for electric vehicle batteries is mined. Thus, mining engineering became an early engineering career option.

Mining engineering courses include beneficiation processes, chemical engineering of ore, mine construction, mine safety, rock mechanics, and the like. Most states in the USA had a mining engineering function as part of their education offerings around 1900. In 2022, mining engineering is offered by a limited number of engineering schools, but it is still a critically needed profession. We still need mines for metals, and in 2022 we need new mines and processes for substances like lithium and rare-earth elements.

Figure 32   *This house is being built with roof shingles that are each a solar cell that generates electricity. They can supply the energy need of this house with excess sold to the local utility company. Mining engineers are greatly needed to develop ways to extract the minerals from the earth to power solar cells, EV batteries and magnets for motors.*

Aerospace Engineering

 Selected schools offer degrees in aerospace engineering to educate and train people to design the rockets and spacecraft for space exploration, as well as equipment for conventional aircraft. Students in this engineering discipline need extra courses in physics and math to learn how to program trajectories of space vehicles or weapons. They have to take extra courses in fluid mechanics to understand propulsion kinetics. Of course, astronomy study is needed to know what's out there in space.

In the USA, space exploration has been transferred from a mostly federal government function to private industry space travel and exploration firms. Space travel is evolving into a tourist industry, but mining on the moon or other planets may become a necessity. Thus, aerospace engineers will be a worldwide need.

Figure 33 *This is a typical day at the beach in 1950 when the world population was less than two billion. The world population was over eight billion when this book was written. Engineering must be employed to deal with ever-increasing population.*

Biomedical/biomechanical Engineering

As previously mentioned, healthcare is the largest industry in the USA in 2022, and it is near the top in many other countries. The pandemic of 2020 served to increase the need for engineers who are trained to engineer needed products and devices related to the human body. For example, a firm was formed in 2022 to develop a device like an earbud to monitor certain aspects of brain activity to prevent seizures. Such a device requires a lot of engineering. Prosthetic devices also require a lot of engineering as do mobility devices such as wheelchairs. Some industries are researching the use of computer-controlled exo-skeletons to assist in lifting and walking. In my specialty, tribology, the field of biotribology had developed to address friction and wear in replacement joints. They now offer hips, knees, and shoulders in many hospitals. "Bio" engineers design and test these devices.

Students in bioengineering have to learn all the things needed to design and build things as well as biological processes, how the human body functions, the many parts of a body (anatomy) and the many terms used to describe body components. Companies that make medical devices hire biomechanical and biomedical engineers to develop products. Biomechanical engineers tend to work on prosthetic devices and things having to do with mobility, while biomedical engineers may design stents or other devices such as a defibrillator to engineer solutions for internal organs that need help.

Most research hospitals employ "bio" engineers and what they do and work on grows in scope each year. The first hip joint replacement was in 1950's. In 2022 joint replacement in the USA was a ten-billion plus dollars per year industry. Bioengineering is a growing career field. I have tribology colleagues who now work in hospital labs rather than engineering school labs.

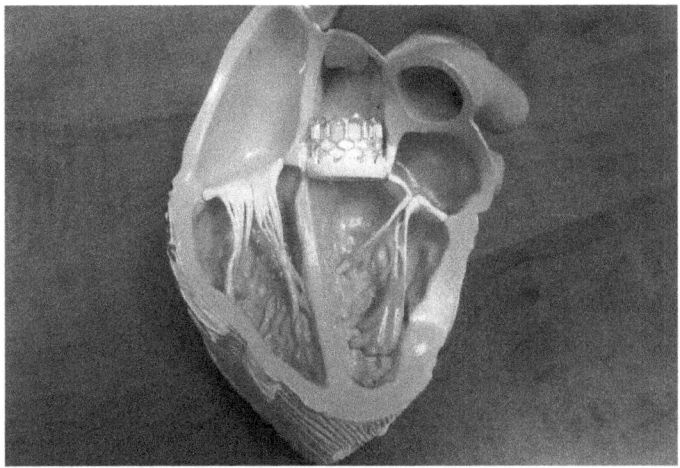

Figure 34 Biomedical engineers designed this artificial heart valve.

Why I chose Biomedical Engineering as a major in college:

Throughout my high school years, I knew I had a natural aptitude for both math and science. I also had a strong interest in medicine. However, I was not sure whether I wanted to pursue a career as a physician.

Biomedical engineering was the perfect fit; it comprised both engineering and medicine. In fact, one third of graduates from biomedical engineering went to medical school. I decided to pursue medical school after taking a physiology class designed for biomedical engineers.

After medical school, I completed a residency in general surgery, followed by fellowships in Surgical Oncology and Surgical Endocrinology. My background in engineering was key to my overall success. Biomedical engineering taught me how to think, which set me apart from other physicians. I often analyzed the materials and tools used in surgery. I also taught myself how to perform statistical analyses which was essential to performing medical research resulting in numerous publications.

I have always wanted to pursue design of medical instruments, but had no time. Recently, Texas A&M has opened a medical school for "physicianeers" where students earn both an MD and a master's degree in

engineering. I am looking forward to seeing how graduates from this program tackle some of health care's greatest challenges.

150

Christine Landry MD, FACS
B.S. Biomedical Engineering 1997
Texas A&M University
M.D. 2003
Texas Tech University Health Sciences Center

Petroleum Engineering

Like some of the other engineering specialties, not every school offers degrees in petroleum engineering, but for years this engineering discipline has commonly scored near the highest in engineers' starting salaries. Petroleum engineers learn how to find and deal with the fluid that has more value than anything else on the planet. We need petroleum for the gasoline to power vehicles, to propel our trains, to fly our airplanes, to pave our highways. Products from petroleum are ubiquitous.

Petroleum engineers have to take chemical engineering courses to understand the chemistry of petroleum. They take courses on how to convert mineral oil to saleable products, the geology of deposits, techniques for "mining" deposits and there are many design-related things to learn about exploration and processing. Details on required courses are in Appendix 1.

Petroleum engineers are hired by oil and gas producers and by companies that make drilling and exploration equipment. This engineering discipline is ever-increasing in importance as demand for sources of energy increases.

Summary

There are probably engineering schools that offer degrees in fields other than the ones that we listed, but that is to be expected. Engineering is a very broad career path with many specialties. A person considering engineering as a career should ponder available disciplines. If it sounds too perplexing, a young person could enroll in one of the big three (Electrical, Mechanical, Civil), and he or she will soon learn more about the more specialized career options in engineering. You will soon learn which courses you like and which you like not-so-much. Try to pick an engineering discipline that you like and do well in. That is the best fit.

Chapter 8 Career Aids

Getting an engineering degree will allow you to call yourself an engineer, but like any field, you have to learn how to practice engineering. I'll never forget my move from the automotive industry (where I was a co-op student for five years) to the chemical process industry. I was hired into a metallurgical lab and I knew how to do metallurgy investigations, but I had no idea what kinds of jobs, projects and activities lay ahead at the huge plant where my job was located. At the time, it consisted of hundreds of buildings and over 30,000 employees at that site. I was rather overwhelmed by the size.

Fortunately, my employer pretty much knew that new employees can be easily overwhelmed and they gave new engineers a "mentor" for about two years. They called these first two years orientation. This was a huge career aid. Then I learned that I had to start reading the literature for engineers in the chemical process industries; then I joined the appropriate technical societies and then learned how to network in this new-to-me industry. The chemical process industry was much different from the automotive industry. In the latter, we mostly made parts, assemblies and machines. In the chemical process industry, the final product could be a tank car of a chemical.

Fortunately, there are career aids that take over after college. These aids help to convert an engineering graduate to a practicing engineer.

It is the purpose of this chapter to discuss the career aids that are available to all types of engineers. The chapter objective is to let young people considering a career in engineering know that help is out there after school. You are not thrown to the wolves. Career aids like orientation, literature, technical societies, continuing education and networking are out there to help you transition from student to practicing engineer.

Orientation

Most companies who hire engineers realize that problem solving in college is much different than on the factory floor. They usually have an orientation for newly hired engineers which may be formal and structured or not so formal and structured. What usually helps the most in getting acclimated to a new company is to be assigned a senior engineer who will show you the ropes. In the automotive job I was assigned a mentor who gave me assignments and showed me how to do the assignments in the way that the company wanted them done. In manufacturing, everything is documented. There are company terms for procedural steps; there are expected records, expected behaviors, expected results.

A good mentor is a person who is high enough in ranking so that you will not compete with him or her in job performance. For example, if you are given a mentor who only has been with the company a year or two, you will likely be competing with this person and thus his or her mentoring may be limited. At my automotive job, my mentor had been working in the specialty (die casting) for more than twenty years. I certainly was not competing with him.

When I started my career in the chemical process industry, I was given a mentor who had been doing his job (welding engineer) for thirty years. Again, I certainly was not competing with him and he was extremely helpful in all matters. He was sort of training me to take his place. He did not withhold information.

Some companies are too small to have an assigned mentor, but you will need orientation anyways. You will need to learn how the company works. One time, I found myself dealing a lot with an analytical chemistry organization. It was even bigger than the engineering division where I worked. I felt that I needed to learn more about what they did and how they functioned. I mentioned this to my boss and he arranged for me to spend a month in this organization working under a senior chemist. I finally learned how to deal effectively with this organization and I established helpful contacts in that organization that lasted my career with the company.

Technical Societies

Whatever the technical field, there is probably a "society" to bring together people in a technical field, to have conferences to share what they have learned and usually to publish research in that area. Most technical societies reach out to students. They have special events, membership fees, scholarships, internships, even summer camps. Technical societies are also a great source of career information.

I joined the metallurgy technical society (ASM) while I was still a student. At that time, there was a robust local chapter in my hometown. They held monthly meetings with an invited technical speaker and I used to attend them during my work blocks in the automotive factory. Years later I became chair of the local chapter and later became a fellow in the society. For those who are not familiar with the term "Fellow", it means that you distinguished yourself in your field and your peers are giving you recognition for this. You are elected/selected by a Fellows committee.

I used to go to erosion conferences at Cambridge University in the UK and being a "Fellow" was a big deal in the ancient colleges. Cambridge University is composed of about forty colleges. Each has their own beautiful campus and center courtyard. If you are a "Fellow" at one of the colleges, you can walk on the lawn in the courtyard. All others are banned (except for the people who mow the lawn). Sir Isaac Newton was a Fellow at Trinity College (c.1700). I later became a Fellow in other engineering societies and I am still active in four technical societies.

Technical society participation should be part of your training and education in your field. Most have a student membership fee that is small and any student is always welcome at each chapter meeting where the cost is that of a meal. Most technical societies meet in restaurants.

Some societies have student chapters at schools and they meet in school facilities.

Some technical societies for the more popular types of engineering are:

Type of Engineering	Related technical society
Electrical	IEEE- Institute of Electrical and Electricians
Mechanical	ASME- American Society of Mechanical Engineers
Chemical	AIChE- American Institute of Chemical Engineers
Industrial	SME- Society of Manufacturing Engineers
Materials	ASM International
Civil	ASC- American Society of Civil Engineers
All	PE Society
All	ASTM International

The latter (ASTM International) is different from the other technical societies in that they are concerned with development of international standards for products, processes and procedures that apply to all fields of engineering. They also welcome students.

Standards apply to engineering because engineers design and build things and this requires metal and plastic shapes, chemicals, fasteners – many components. International standards determine what is out there to use in design and building. For example, a civil engineer charged with building a building has to specify sizes of steel columns, beams, concrete, fasteners, windows etc. However, there are a limited number of sizes available in steel shapes, all bolts are only available in certain sizes and shapes, and concrete mixtures are only available in several formulations. International standards or country-wide standards determine what is available to use in building a building. Standards organizations establish what is available to build something: what size steel columns are available, what size steel beams are available, what types of steel are available and on and on. Standards are necessary for world order and civilization in general. Steel mills can make most any size beam or column, but since the equipment to make a single steel shape, like an I-beam, may cost tens of millions of dollars, they are not going to make bespoke shapes. They make standard shapes. Lumber mills do similar; they only offer certain shapes and sizes of lumber, those determined by relevant standards. Only certain fastener sizes are made; all dimensional lumber shapes had to be made in a particular way; all oils are made to certain viscosity and property standards and thus the companies in the world who make things agree on what they are going to offer in size, shape, function, materials, etc. Engineers must build things with materials and products that reflect worldwide availability.

ASTM has committees on hundreds of engineering subjects. Committees develop the standards and what evolves as a standard must be the consensus of a committee by vote. I got involved in a committee which dealt with my engineering specialty – tribology (friction, wear, erosion). Members of a committee meet formally twice a year and decide on suitable standard projects. They hold symposia on technical issues and the product of the committees are standards in a particular area. The company that I currently work for (my son's company) performs tests in friction and wear for clients using mostly standard tests.

The other technical societies that I belong to do not develop standards, but their conferences and publications serve as the body of knowledge in their particular field. For example, the materials society publishes relevant information on engineering materials, the lubrication society is where the lubrication engineers agree on what type of lubricants to make for general use, the bridge society brings together engineers involved in building bridges and so on. Technical societies link engineers with colleagues doing like work.

Figure 35 *Fourteen hundred lubrication engineers from 37 countries are waiting to get into a conference lunch event. Engineering society conferences are for sharing earnings, for getting new ideas from others, and for networking. These conferences let you know what is happening in your field of engineering.*

Continuing Education

An engineer's education and training does not end with getting your degree or in passing a licensing test. Technology evolves and what I learned on how to design a machine is much different from the way things are done now. For example, we designed a machine on a drawing board with an appropriately-sized piece of blank paper. Drafting papers (and drawing boards) are almost nonexistent in my city. Most designing of machines and equipment now is done on a computer screen. Computers as we now know them did not exist in 1961 when I got my first engineering degree. I had to learn how to use them. I had to learn many new things over the years to do my job. "Continuous education" is the term used to describe what common sense tells professionals what they must do to keep up, etc. with the technology of their field.

Evening school is a convenient way to continue your engineering education. All cities with colleges and universities have evening school or day school that can provide a young engineer with training and education that may be needed for a new job situation. When I started with the chemical process industry I soon learned that my college education was deficient in corrosion engineering and in polymer engineering. I enrolled in evening school to take classes in both areas. It was just what I needed since my new employer had lots of corrosion problems and they were a manufacturer of plastics. When PCs came along, I went to courses on them at local high school evening programs. (I bought by first PC in 1983, and I paid $4300 for one that did not have upper and lower case on the keyboard). Overall, what I learned in evening courses was invaluable for my career.

If you have an engineering career that requires state licensing, you may have to document your continuing education in order to keep your PE license. For example, in New York State you need 39 hours of continuing-education credits in a three-year period to keep your license. Courses need to be state approved. Each state in the USA will have different requirements. Some states may allow on-line courses.

Networking

Networking is the millennial term for "schmoozing with peers". When I started attending technical society meetings, I would socialize with the other attendees and after a year or two I knew all of the materials engineers in the city, where they worked, what they did and what it was like to work at each company. They call that networking in 2020, but it is the same as I did at local technical society meetings.

I now get notices of networking events from the overarching engineering society in my city. They have more formal events now. Social gatherings with peers can be helpful to a career, and participation is encouraged. They are a bit more formal, but the intent is the same: get to know what your engineering peers are doing.

Networking done right requires that you continually reach out to others that you do not know. If you go to a networking event and sit with people you know or work with you will not learn about what is happening in your field at other companies. A valuable benefit of networking is knowing if your current job is better, equal or lesser to peer jobs at other companies. For example, I learned after a year of two of networking that my current position was better than if I had a comparable materials engineering position at the other companies in town.

Networking can be a great tool to assess your current job situation. You will learn if things are really better at XYZ Corporation. Conversely, you may learn that things may be better at XYZ Corporation for what you do; you may want to consider a job change.

Networking is a great way to find out what's new in your field. What are your technical colleagues working on? Is there something that they do that you should adopt? You can also learn about technical capabilities. Heat treating is a key part of metallurgy. When I went to local technical meetings, I learned about facilities available in the area. This information allowed me to solve some significant materials problems at my company.

Recently I attended a biannual research conference at a New England private college. These conferences have been held for decades, and they bring together researchers from all over the world to share current work. This particular conference requires living in a college dorm for a week and eating meals together. I always would sit with new-to-me people at every meal. Many of the attendees were PhD students or "Post Docs" (post docs have finished all degree requirements, but stay on for a while at their school or work temporarily at some research facility before seeking full time employment) and eating meals with young people let me know what lies in the future in my field and for my company.

In summary, to be good in any career you have to be commensurate with current practices and technology. In engineering, there are plentiful aids to stay current and improve your "technical content". The more you know, the more you are worth. This is a cliché, but it is true. Throughout my life I noted that those who really know their field are the most rewarded. However, more importantly, people who are on top of their field seem to be happier. Knowing your field as well as possible gives you the self-confidence to keep progressing in your field.

All of the "aids" that are mentioned in this chapter are effective in enhancing a career in engineering. They are usually pain free. Meeting others, meeting more peers, often produces unanticipated positive results. For example, many engineers that I met in technical society activities ended up as lifelong friends. And more engineering friends are helpful to any engineering career.

Figure 36 *Engineers design the buildings in a city center, erosion control on waterways, bridges, roads, drainage; just about everything that makes a city happen is designed by engineers.*

Chapter 9 Life as an Engineer

Start of Career

What is it like to be an engineer? A first job in engineering will probably be a bit scary. You have been hired to bring new skills to a company and produce savings and benefits to the company that amount to at least twice your annual salary. You are expected to be immediately productive. However, you may be overwhelmed by your new environment.

Starting a new career is like starting anything new. When my wife of fifty years died, I had to learn how to cook, clean, and run a household. It takes effort and there will be some failures. This chapter is about what the various phases of an engineering career may look like.

The start of my engineering career went well because I had good orientation programs and because my early assignments were assigned to me by others. I was given projects to work on by my mentor or boss. These were actually the halcyon years. I later learned that I had to come up with innovative projects, rather than have them assigned to me. However, I put that in mid-career category. Conceiving of new products, processes, machines etc. is part of senior engineering.

On the subject of senior engineering, most large companies that hire engineers usually categorize you in some way. Colleges do something similar with faculty. You may start as associate professor, then become professor, then get a chair and so on. In my chemical process industry job, I started as a code 43 engineer, the next step was a code 45 senior engineer, then code 47 technical associate, then code 49 associate. Each code level reflected an increased salary range. Code 49 associates often had higher salaries than their boss. This was the case with a company that had a dual path; you could grow in your technical field, or grow in company management. This recognizes that top technical talent may make poor bosses and some people excel at handling people problems and the like associated with being a manager. Determining if a company has a dual path to career growth is a very important job selection consideration. If you love the design aspect of engineering you do not have to give it up to grow in your career. You do not have to become a boss to earn more money.

My early days in engineering were the most fun. I was always learning new things; I started with other young engineers and we bonded. We did things together. We became close friends. As an aside, I recently visited an engineer that I worked with in my early days. He left the lab where we worked together in 1968. He lives on the water in an historic seaport in New England and I have been visiting him every few years ever since. This is the kind of bonding that can occur in working with other engineers in a group or team mode. The social aspects of working with others can be a huge career benefit.

A typical early year project would be something like investigating a sulfuric acid leak in the synthetic chemicals plant. I would go to the site, pinpoint the leak and make recommendations on materials and equipment to solve the leakage problem. We would write up a technical memorandum with the lab tests results, conclusions and recommendations. Writing reports was standard practice for the type of work I was doing. Most projects pertained to a production problem and the related report contained the action items. My documentation of what was done on a project heavily influenced my performance evaluation. In fact, in all areas of engineering, communication skills are essential. You must write clearly and concisely. You must document your work. You must become effective in making oral presentations to peers and management.

A second phase in my early years in the chemical process industry was to initiate development activities. For example, investigate the use of a new material for application in production processes. For years, only one metal could be used for tanks holding bleach solutions. It was very expensive, available from only one company, and it was very difficult to machine and fabricate. Finding and testing a suitable replacement was a typical development project. Of course, this type of project would involve laboratory testing to ensure that a replacement material would have the required corrosion resistance.

My personal life in my early career was busy with night school, raising three sons, teaching them how to ski and sail. I was working my way up the ladder in my local technical society and I took on responsibilities in my church and community. It was a super busy time in my life and one of the happiest. I never had time to wonder if my career was on track. I loved my job and the company that I worked for.

Mid-career

Most of my engineering peers considered their early career to be from graduation from engineering school to about thirty years of age (the first ten years). Mid-career was from about thirty to forty-five years or the next fifteen years. This is usually the determining part of your career. Here is where you will do your best work. Here is where you will likely determine your career outcome.

In my case, I evolved from factory floor problem solver to researcher in tribology and tribomaterials. My development projects often became fundamental research. I became heavily involved with several technical societies and started to publish papers in their journals on a regular basis. I taught engineering materials at two local colleges (evening school) and my teaching experience led to writing a teaching textbook. My textbook led to two U.S. National Science Foundation missions to other countries to promote technology in tribology. I ended this mid-career phase as my company's expert in the field of tribology and I had a national and international reputation in this area.

My personal life in mid-career was even busier than my early career life. I had to work on career guidance for my three sons. I moved up to a large cruising-type sailboat and I was taking the family "out west" to ski. We lived sort of a storybook type of life. We had sufficient income so that my wife did not work outside the home and we had all of the worldly possessions that young people perceived that they need. I was elected fellow in my main technical society. My employer was healthy and research funding for my work was ample. These were the wonder years of my career.

Figure 37   *An engineer's life is often full of adventure and travel; sometimes travel is part of his or her job. Sometimes it is part of the curiosity that is usually present in engineers. This is Rio DiJainero, Brazil*

Career Plateau

From about forty-five to fifty-five years of age is what I term a career plateau. I achieved all of my technical goals and plateaued to one publication in a peer-research technical journal per year. This was sort of my career plateau. I did not want to move into management and I decided to do fundamental research on engineering material problems that were a limiting factor in my company. For example, what is limiting our speed to produce polyester film? What can be done to prevent open-circuit problems in customer cameras by fretting corrosion produced by motions caused by carrying cameras on your person?

There was no shortage of significant challenges to work on. I also worked on new product proposals such as lithium batteries. In every project throughout my career I had to justify the work with savings that would pay for the work many times over. This is a very, very important point in considering engineering as career. You must justify your salary by savings or new product sales. Engineers are supposed to be "free" to a company. If your annual salary is $100,000 you will need to decrease manufacturing costs by several times that per year. At least that is the accepted concept in large established companies. If you work at a start-up, your salary may be paid by developing a product that produces profits that were several times your salary.

Engineers pay for themselves.

Career Decline

From about age fifty-five to sixty-five many professionals have what is termed a career decline. Although it is against the law, all companies in the United States discriminate against people above fifty-five years of age. If you end up in the job market at age fifty-five or older, it is likely that you will not get a permanent job if you are in the market. You are likely to have some health issues at this age and employers that pay health benefits are reluctant to add employees that may require significant health costs. In addition, your salary requirement is probably high at this point because of all of your experience. So, employers prefer to hire young people at lower salary and health costs. Also, training a person at fifty-five may mean losing that training at your sixty-five-year retirement. Thus, changing jobs in engineering after age fifty-five can be troublesome. If possible, try to end your career with a long-time employer.

My "past sixty-five" work life is about the same as my mid-career work life. I spend most of my time on tribology research. My personal life is now one of two to four technical conferences per year between health issues. I also travel more.

Summary

I am quite old; so old I try not to think about it. However, in my sixty plus years as an engineer, I can truthfully say that I never encountered an unhappy engineer. We engineers have all had up and down cycles in funding and job availability, but engineers are usually content with their selection of engineering as a career. Many engineers gravitate to other fields, but those that stay in hard-core engineering, like what they do and they usually have an action-packed private life style. When I belonged to a yacht club, ski club and country club, many other members of these clubs were engineers.

By definition, an engineer is an innovator (a new design), a risk taker (the design may not work) and a problem solver (must make a design work). Engineers usually have a life that is as full and rewarding as life can be. Lots of my engineer colleagues have died and invariably their family will note the person's most significant engineering contribution. Most engineers will have some accomplishment that provided the "meaning" of his or her life. My engineer ski-partner was proudest about a computer chip-wafering blade that he invented. The welding engineer in our lab thought that the huge static mixers that he designed and built were his biggest contributions; they aerated municipal sewage. The engineer that sat next to me for years considered his formulation and processing of a certain partially-stabilized zirconia to be his contribution to the world. I will probably designate my technical books as my life's contribution to the world.

Engineers usually live a comfortable active life and they believe that their life helped humanity in some way. This was their "meaning of life". We try to leave things a bit better than we found them.

Chapter 10 Future Engineering Challenges

If the person-on-the-street was asked in 2020 what are today's top three issues; most replies would include: 1. the environment. 2. The economy. 3. Peace and order in society. Worldwide media have painted a doomsday scenario when it comes to the environment: The oceans are rising, the forests are disappearing, and the world is getting too hot to live in.

The economy has always cycled during my life, but in 2022 we are seeing something that I have never seen in my lifetime: in the USA, hardly any new cars are available. This is new, this is very serious. A global economy has produced the disappearance of products that were always taken for granted in the USA.

Societal issues are present in every country, but what I have not witnessed in my lifetime is the vilification of security personnel. Police in the USA can get accused of murder if they make a wrong split-second judgement. In addition, social media has become an addiction for many.

How can engineers help these issues in the future? What are society's biggest needs that can be met by engineers? What will engineering look like in 2030 or 2050?

It is the purpose of this chapter to predict the role of engineers in a post 2022 world. The chapter objective is to give a person considering a career in engineering an idea of what lies ahead if he or she elects engineering as a career. The chapter format is environmental issues, economy issues and societal needs.

Environmental Issues

Engineers pretty much are solely responsible for fixing the planet's environmental issues. We have to reduce carbon dioxide in the air, we have to create clean energy, we have to armor the waterfronts of coastal cities, we have to stop asteroid collisions etc. Indigenous people in the Americas developed dams and canals to bring water to dry areas thousands of years ago. This kind of thing needs to be done again. Many parts of the world are losing their rain, their water for crops. One technique that I witnessed that worked for hundreds of years in Bermuda is the capture of all rain that falls.

Bermuda has no fresh water. It is the peak of a volcano. All roofs are made from a stone tile and all water that falls on roofs goes to cisterns. All rains that fall on public places go into lined catch basins and other reservoirs. Every drop of rain that falls on the island is trapped.

Every year in western USA wild fires burn down forests then big rains come and wash the soil away. Engineers could design holding systems to retain monsoon rainfalls for the dry season. This could be done worldwide.

There is more water area than land on the planet, but seawater is useless because of its salt content. Engineers could design an evaporation system for removing the salt that does not require artificial energy.

There is enough hydrogen in sea water to power every vehicle in the planet, but we lack a cost-effective way to get the hydrogen atoms away from their oxygen atom. Some brilliant engineer can conceive of a way of doing this.

Sewage is a costly problem in every country. It contaminates the drinking and potable water. Trees and other plants love sewage. It is their "chocolate candy". Trees will send their roots just about anywhere to tap into sewage and drink it.

Engineers could design cleaning forests full of vegetation that clean sewage as well as generate needed oxygen. People have been using human waste for fertilizer for centuries. This is the same concept, but using trees as plants that are not to be eaten as the cleansing part of the system.

Burning fuels to heat and cool could be significantly reduced by engineering zero-heat houses and other buildings. Engineers have demonstrated this concept. It has been shown to work for decades. You simply insolate inside from outside enough so that body and machinery waste heat can heat a living space and you cool the same living space using the earth below which is always about 10 degrees C. It can cool. However, engineers have to design the system.

Global warming is mostly the product of burning fossil fuels for transportation. Incredibly, prior to the steam and internal combustion engine, we did not burn any fossil fuels to move goods and services worldwide. We sailed across oceans. We used beasts of burden to move large objects. Engineers could design a fossil-fuel-free goods transportation system for the future.

Airplanes emit bad things into the atmosphere. A new aircraft fuel (possibly hydrogen) needs to be engineered.

There are many people to feed. The ocean is most of our planet and it is full of living things that can be eaten. The ocean could be engineered as a source of food. For thousands of years, indigenous people in various locations lived off of shell fish and other fish. There is no limit to the possible ways that engineering could be used to fix our environmental dilemma. Currently there are very few engineers working on existential problems. That will have to change in the future.

Figure 38 *This is what is inside of the "battery" of a typical electric vehicle. Hundreds of individual batteries are welded together to supply enough energy for about 200 miles of driving. A challenge for future engineers is to make a single cell to do the job. Welding hundreds of little batteries together creates hundreds of opportunities for something to go wrong. Also these battery packs of individual cells are very heavy. Weight reduction is another engineering challenge.*

Economic Issues

In 2022, the Covid-19 pandemic that started in 2019 is still making many unavailable items that people need. In the USA, shortages and unavailable items probably have a root cause in insufficient staff to supply, make, serve or deliver items. There are not enough people to bring goods and services to the population. There are not enough school teachers. There are not enough school bus drivers, there are not enough sports coaches. The available labor force is shrinking.

Engineers can help alleviate labor shortages by designing systems that can do work without an active person being involved. For example, when I worked in the photographic film industry, each film product was made on a different film support. At the facility where I worked, engineers designed and built a gigantic warehouse for rolls of different film supports. The warehouse could hold maybe 10,000 rolls of film. Each weighed 2000 pounds and was at least five feet long and two feet in diameter. Engineers developed the racking and retrieval hardware so that a gigantic warehouse full of rolls could be computer controlled such that a single roll could be retrieved and delivered to a specific spot in minutes by keystrokes on a computer keyboard. There were no people involved in storing and retrieving rolls of film.

In 2022, on-line commerce works that way, but labor is still needed to do home delivery. Engineers are currently designing autonomous vehicles to do the delivery. They are in place in some areas of the world in 2022 and their use is growing.

Autonomous school buses would solve the bus driver shortage problem. Autonomous farm vehicles are helping solve farm labor shortages. Autonomous off-road trucks drive to the bottom of pit mines for iron ore. Drones can apply pesticides to farm fields with reduced labor requirements; autonomous ships may be plying the oceans to deliver goods in the future. Autonomous devices are designed by engineers.

Another significant way that engineers affect the world economy is in engineering materials. Lithium in batteries has made an incredible effect on the usability of batteries to power everything from phones to trucks. In fact, the battery industry is ready for a new cathode and anode material that takes less from the environment. Engineers are needed to do the development.

Historically, human civilization correlated with developments in engineering materials. Copper technology brought a huge increase in human ability to make things. The technology of iron brought another major change in economies of the world and of course, the development of steels by metallurgical engineers in 1850 or so allowed the manufacturing economy that brought incredible living capabilities to much of the world. Materials engineers brought us all of the things that we now need and expect.

Homo sapiens had to live in caves and hunt with stone spears for thousands of years because they did not have metal tools. Engineering materials are that transformative. Steel allowed everything to change. Semiconductors allowed electronics to be everywhere. This technology occurred within my lifetime. My master's thesis involved growing a large single crystal of pure iron by a process called zone refining. This crystal-growing process and relative technology is now the basis for growing silicon single crystals that are the basis for the computer chips that in 2022 control everything. This material capability did not exist in commercial form in 1960. Thus, engineering advances in the last sixty years or so has transformed not just the economy but everything. Engineering is that transformative.

Society Issues

When I started working as a design engineer in the US automotive industry (around 1960) my employer made me wear a shirt with a collar and tie (with a tie pin), join the local chamber of commerce and give blood every time that the Red Cross asked for it in a blood drive. I was expected to be a "model citizen". I balked at giving blood since I faint at the sight of blood. My boss told me that he would take me to the cafeteria site and hold my hand, but I had to give blood. I closed my eyes through the whole procedure, but that was the way that things were. Engineers were expected to be model citizens in all matters.

Incredibly, the situation is still the same. I have worked with other engineers from all over the world and I have to say that all engineers that I met and worked with tried to be model citizens. They all participated in their civilizations by informing themselves of their situations and making societal decisions based upon ethics and morality. I had a Russian engineer friend that fled the USSR when it was breaking apart. He was a genius and loved his mother country and his heritage, but knew that it was time to leave. Fellow scientists in the USSR were selling furniture to survive. He and his wife with similar engineering talent, (she was a movie producer in the USSR), built an American dream life in the USA each starting with a suitcase with a few changes of clothes and a few belongings. They taught themselves English and did whatever they could to earn rent money. Eventually Nikolai got an engineering job and they went on to own the traditional home in the suburbs and a vacation home in a resort area. They started this at age fifty-five. They became US citizens and did what was needed to be good citizens of their new country.

Unfortunately, societies seem to have a tendency to swerve wayward over time. The greatest ancient societies in recorded history lasted for thousands of years (Egyptian, Greece, Roman etc.), but all eventually faltered. The people still exist, but their societies no longer cover large portions of the inhabited earth. What killed the great societies? They stopped providing their populace with basic human needs: a job (food, clothing, and shelter), peace, and protection.

Humans are born with self-preservation instincts that enable them to find food, clothing and shelter. They started banding, living in groups for peace and protection. This is where engineers started to shine. The engineers in a group of several families emerged and designed essentials for improving their situation. The earliest "city" uncovered in Mesopotamia (circa 9000 B.C.) had intricately designed buildings with roof-top entrances. The rooftop entrances likely had to do with protection. The houses were interconnected indicating a peaceful situation between residents. This civilization lasted for an estimated thousand years.

It is incumbent upon people who are gifted with abilities commensurate with engineering to also do their part in making society a place where everyone can have a job, protection and peace.

Engineers seem to have a gene-related motivation to do what is right for the society in which they live.

Figure 39 *In summary, please consider joining these*
students in engineering school

Epilogue

The purpose of this book is to describe what engineering is all about, what it looks like, what life as an engineer may look like. Choosing a career is an important part of life and a young person often does not have the "data" to compare many careers, but a career decision needs to be made at least in high school, since many schools offer college courses in junior and senior years in high school.

I wish that all high schools made career guidance a priority. It was when I was in high school, but things are very different now. In much of the world, manufacturing is no longer the mainstay of local economies. In the USA, healthcare has taken over as the most likely place to work. However, engineering is an integral part of healthcare. Health care products need to be manufactured. Factories are needed for that and engineers are needed to design factories. Someone has to deal with design and construction of medical devices; engineers do that. Joint replacement is a large business in many countries including the USA. Design and building of ambulatory devices, like wheel chairs and exo-skeletons is part of engineering. And of course, robotics, which has always been a part of engineering, is now a critical part of the health care industry. My last operation was performed by the DaVinci robot with the surgeon at a console not at the operating table (but not in the cafeteria).

Overall, many careers include some form of engineering and it can be used as the basis of another career since engineering courses teach fundamentals like strength of materials, that apply to all solid substances. Fluid dynamics applies to the behavior of all liquids. The science of engineering materials really applies to many careers. What is a plastic? What is a ceramic? What is a cermet? Just about any career needs to know the basis of engineering materials.

What do engineers do in their daily lives? We tried to answer this with personal examples, but what an engineer does depends on the company that the engineer works for. Most engineers spend the bulk of their time at work solving problems for their employer. In fact, engineering educations generally teach problem solving coupled with how to communicate how a problem was solved. Communications is a big part of engineering.

Writing reports on how to do something is part of the business. Making oral and on-line presentations is part of the work. Researching subjects is also part of the business. What was previously done is usually the starting point of any new study or project. Dealing with others is also part of the business. Teams are often used to deal with problems and design issues. Engineers have to be able to work with others.

Chapter 4 listed some of the great engineering achievements in history, but the cell phone that billions of people carry on their person is testimony to an incredible engineering achievement. The ability to use GPS in every trip to the doctor's office or to an across town location is an incredible engineering achievement. You are using satellites in space to tell you where to make your next right turn. Every improvement in the status of civilization has been produced by some engineering achievement. The wheel was the start of vehicles; the discovery of copper brought metal tools; the development of iron technology brought even better tools; better tools allowed ships which spread civilizations everywhere. Incremental engineering achievements over the centuries brought humans from hunter/gatherers to what we are today.

We addressed matching your interests with engineering in chapter five. Engineers have always been called "handy" in American jargon. Engineers tend to be able to work with their hands, to be able to build things. Not everybody has this trait. My three sons spent their preschool years building forts and the like in our backyard. They were all builders. In high school a huge skateboard ramp showed up in our backyard, then motorcycles in various stages of construction. These are clear indications of having an inclination towards engineering.

Sometimes just the order and analytical ability of a person is all that makes him or her a good candidate for an engineering career. Engineering teaches one how to solve problems and documents what was done. Some people have this as an innate ability. Engineering would be a compatible career for them.

The school requirements for engineering vary with the type of engineering, but basically engineering school tells students how things work. All engineers get courses in math, use of computers, physics, chemistry, fluid mechanics, and strength of materials, design concepts – all of the basics needed to be a technical problem solver.

The first two years are the same in just about any engineering school. The junior and senior year courses are determined by your choice of specialty in engineering.

Cooperative engineering schools often make work blocks part of the learning process by reports on work block projects. Some colleges even require a thesis project for a BS degree. The more hands-on a program is, the better the engineering (an opinion).

Choosing a type of engineering is the same as choosing a type of medicine to practice as an MD. The human body has so many organs to deal with. A doctor cannot be an expert on all. In addition, medical specialists like a particular body area. Thank goodness! I have plumbing problems and I am thankful that urologists specialize in body plumbing.

Selecting electrical engineering over the other types will mean that you will likely work on controls or power generation/transmission. You will not be building buildings or designing machines. My brother, the electrical engineer in the family, spent his career designing the small electrical motors in vehicles that control windows, seats and all other sorts of things.

You will need to make a choice on what type of engineering you want to do. Of course, you can change fields, but it has been my experience that once a person selects a type of engineering he or she stays with that choice for a career, like medical specialists.

The career aids that we discussed in Chapter 8 are very real; they are very valuable. You really need to keep abreast of improvements in your field and you really need to develop networking contacts. Technical society participation has made my career quite different than I perceived. I expected to spend my life in a particular factory. Instead, I traveled all over the world giving papers at conferences and participating in activities that I never even knew about as a "newbie" engineer. Career aids are to be embraced and enhanced by participation.

Life as an engineer is what you make it. However, engineering salaries in most countries are good enough to allow what is termed a comfortable life. I bought a new house the first year out of graduate school. I built two customs homes after the first one. My wife never had to get an "outside" job and life was characterized by financial stability. We always had enough; we socialized with other engineering families; my coworkers became life-long friends. It was a very good life.

Future engineering challenges really make engineering the number one career choice for those wanting to make a difference with his or her life. In 2022, there are so many existential problems that everybody who can, should go into engineering. Our problems are so great, so numerous, so challenging, that governments and corporations should be giving "sign on" bonuses for starting engineering school.

Global warming can only be solved by engineering advances. Clean energy must be solved by engineering. Our fresh water crisis needs engineering solutions. Even implementation of electric vehicles will require a huge engineering effort. A street of fifty houses cannot charge their electrical vehicles at the same time because of the incredible current load. What are EV users to do? A solution must be engineered.

The picture that we have tried to paint in this epilogue is that engineering is so important for the survival of our current civilization. The global economy has scattered essential industries all over the world and the least interruption in shipping and manufacturing can shut down the planet. Each country on the planet needs to ponder that country's sustainability. When they do, they will find that they need to try to get a very large portion of their young people interested in engineering. Survival of the human species depends on engineering solutions to the limiting problems of our times.

kgb

Appendix

Sample course requirements for various types of engineering

Note: Every college has unique course requirements. However, most engineering colleges require certain basic courses for each type of engineering. A specific engineering school's course requirements can be found on their website. The course requirements in this appendix are taken from websites of various engineering schools, but a significant portion of these course listings come from the Texas A & M University website.

Part of screening potential schools for an engineering degree is evaluation of their course requirements. Some engineering schools may have a thesis or project requirement. Course evaluation and degree requirements are a necessary part of exploring engineering as a career.

Petroleum Engineering BS

First year	Fall	Spring
	Chemistry	Physics
	Chemistry Lab	Engr. Math
	English	Mechanics
	Computation	Core course
	Engr. Math	Chemistry
	Core course	

2nd year	Fall	Spring
	Geology	Mechanics
	Differential Equations	Engr. Math
	Intro to Drilling Systems	Statics
	Petrophysics	Electricity
	Thermodynamics	Physics

3rd year	Fall	Spring
	Geology of Petroleum	Formation Evaluation
	Data Analytics	Reservoir Engineering
	Numerical Methods	Well Testing
	Reservoir Fluids	Petro Production Systems
	Technical Presentations	Petro Project

4th year	Fall	Spring
	Summer Practice	Integrated Asset Development
	Drilling Engineering	Core Course
	Reservoir Simulation	Ethics in Engineering
	Production Engr.	Technical Elective
	Technical Presentations	Student Project

Mechanical Engineering BS

First year   <u>Fall</u>   <u>Spring</u>

Fall	Spring
Chemistry	Physics
Chemistry Lab	Engr. Math
English	Mechanics
Engr. Lab	Chemistry
Core Course	Core Course

2nd year   <u>Fall</u>   <u>Spring</u>

Fall	Spring
Physics	Electrical Engineering
Engr. Math	Differential Equations
Geometric Modeling	Materials & Manufacturing
Engr. Mechanics	Thermodynamics
Electricity	Mechanical Measurements
Statistics	Core Course

3rd year   <u>Fall</u>   <u>Spring</u>

Fall	Spring
Solid Mechanics	Selection of Materials
Fluid Mechanics	Manufacturing Lab
Fluids Lab	Dynamic Systems & Controls
Engr. Analysis	Systems Lab
Dynamics & Vibrations	Solid Mechanics
Seminar	Heat Transfer
Core Course	Heat Transfer Lab

4th year   <u>Fall</u>   <u>Spring</u>

Fall	Spring
Economic Analysis	Intermediate Design
Mechanical Design	Core Course
Core Course	Technical Elective
Technical Elective	Technical Elective
Technical Elective	Core Course

Industrial Engineering BS

First year	Fall	Spring
	Chemistry	Physics
	Chemistry Lab	Engr. Math
	English	Mechanics
	Lab Computation	Chemistry
	Engr. Math	Core Course
	Core Course	

2nd year	Fall	Spring
	Physics	Programming
	Industrial engr. Design	Informatics
	Engr. Math	Linear Algebra
	Electricity	Statics
	Computer concepts	Manufacturing Processes
	Public Speaking	Statistics

3rd year	Fall	Spring
	Uncertainty Modeling	Operations Research
	Operations Research	Quality Engineering
	Human Systems	System Simulation
	Differential Equations	Production Systems
	Computer Science	Professional Development
	Thermodynamics	

4th year	Fall	Spring
	Core Course	Core Course
	Technical Electives	Technical Electives

Electrical Engineering BS

First year   <u>Fall</u>   <u>Spring</u>
Chemistry   Physics
Chemistry Lab   Engr. Math
English   Mechanics
Computation   Chemistry
Engr. Math   Core Course
Core Course

2nd year   <u>Fall</u>   <u>Spring</u>
Program Design   Electric Circuit Theory
Digital System Design   Differential Equations
Engr. Math   Applied Math
Electricity   Core Course
Physics

3rd year   <u>Fall</u>   <u>Spring</u>
Signals & Systems   Random Signals
Magnetic fields   Electricity Conservation
Electronics   Computer Architecture
Physics   Electronic Properties of Materials
Communication   Technical Elective

4th year   <u>Fall</u>   <u>Spring</u>
Design Lab   Design Lab
Technical Electives   Technical Electives
Professional Development   Core Course
Core Course   Elective

Computer Science BS

First year

Fall	Spring
Chemistry	Physics
Chemistry lab	Engr. Math
English	Engr. Mechanics
Engr. Lab	Chemistry
Engr. Math	Core Curriculum Course

2nd year

Fall	Spring
Intro to Computing	Algorithms
Program Design	Computer Organization
Discrete Structures	Programming Languages
Linear Algebra	Communication
Elective	

3rd year

Fall	Spring
Intro to Computer Systems	Algorithms design
Foundation of Software	Seminar
Statistics	Engr. Math
Core Curriculum Course	Differential Equations

4th year

Fall	Spring
Core curriculum Course	Circuit Design
Computer Elective	Core Curriculum Course
Concentration Elective	Computer Elective
Elective	Concentration Elective

Computer Engineering BS

First year	Fall	Spring
	Chemistry	Physics
	Chemistry Lab	Engr. Math
	English	Mechanics
	Engr. Lab	Chemistry
	Engr. Math	Core Curriculum Course

2nd year	Fall	Spring
	Program Design	Intro Structure/Algorithms
	Digital Systems Design	Discrete structures
	Engr. Math	Electrical circuit theory
	Electricity	Statistics
	Physics	Differential Equations

3rd year	Fall	Spring
	Computer Systems	Software Engineering
	Seminar	Electronics
	Signals and Systems	Integrated Circuit Design
	Math	Integrated Circuit Design
	Communication	

4th year	Fall	Spring
	Senior Design	Senior Design
	Core Curriculum Course	Core Curriculum Course
	Concentration Elective	Concentration Elective
	Engineering Elective	Professional Development

Civil Engineering BS

First year	Fall	Spring
	Chemistry	Physics
	Chemistry Lab	Engr. Math
	English	Mechanics
	Engr. Lab	Chemistry
	Engr. Math	Elective

2nd year	Fall	Spring

2nd year

	Fall	Spring
	Intro to Civil Engr.	Computer Applications
	Statics	Civil Engr. Measurement
	Graphics & design	Mechanics of Materials
	Physics	Fluid Dynamics
	Electricity	Technical Writing

3rd year

	Fall	Spring
	Materials Engineering	Professional Development
	Civil engr. Systems	Differential Equations
	Theory of Structures	Core Course
	Dynamics	Elective

4th year

	Fall	Spring
	Professional Practice	Ethics and Engineering
	Technical Electives	Technical Electives
	Core Course	Core Course
	Concrete Design	Highway Materials

Chemical Engineering BS

First year	Fall	Spring
	Chemistry	Physics
	Chemistry Lab	Mechanics Lab
	English	Mechanics
	Engineering Lab	Chemistry
	Engineering Math	Core Course
	Core Course	

2nd year	Fall	Spring
	Organic Chemistry I	Organic Chemistry II
	Chemical Engineering I	Chemical Thermodynamics
	Physics	Technical Writing
	Electricity	Differential Equations
	Engineering Math	Core Course

3rd year	Fall	Spring
	Fluid Operations	Physical Chemistry
	Numerical Analysis	Heat Transfer
	Chemical Materials	Mass Transfer
	Chemical Thermodynamics II	Kinetics/Process Design
	Science Elective	Core course

4th year	Fall	Spring
	Process Simulation	Plant Design
	Chem. Engr. Lab I	Chem. Engr. Lab II
	Process Dynamics & Control	Process Safety
	Bioprocess Engr.	Major Elective
	Major Course	Core Course

Biomedical Engineering BS

First year   <u>Fall</u>   <u>Spring</u>

Chemistry	Physics
English	Engr. Math
Engr. Lab	Mechanics
Computation	Chemistry
Engr. Math	Core Course
Core Course	

2nd year   <u>Fall</u>   <u>Spring</u>

Pathways in Biomedical Engr.	Bio Applications & Signals
Computing in Biomedical	Medical Device Design
Physics	Organic Chemistry
Engr. Math	Differential Equations
Electricity	Physiology for Engr. II
Physiology for Engr. I	Public Speaking

3rd year   <u>Fall</u>   <u>Spring</u>

Biomedical Electronics	Bio-response to Medical Devices
Instrumentation	Biomaterials II
Biofluid Mechanics	Medical Device Design II
Biomaterials I	Biosolid Mechanics
Statistics	Medical Imaging
Core Course	Core Course

4th year   <u>Fall</u>   <u>Spring</u>

Bio Mass & Energy Transfer	Case Studies
Analysis & Design	Analysis & Design Project
Biomechanics Lab	Core Course
Technical Elective	Technical Elective
Core Course	Technical Elective

Aerospace Engineering BS

First year   <u>Fall</u>   <u>Spring</u>
Chemistry   Physics
Chemistry Lab   Engr. Math
English   Aerospace Mechanics
Engr. Math   Chemistry
Core Curriculum Course

2nd year   <u>Fall</u>   <u>Spring</u>
Introduction to Flight   Aerospace Mechanics
Intro to Aerothermodynamics   Aerospace Mechanics of
Aerospace computation   Materials
Elective   Directed Studies
Core Curriculum Course

3rd year   <u>Fall</u>   <u>Spring</u>
Aero research   Professional Development
Theoretical Aerodynamics   High Speed Aerodynamics
Aero Structural analysis I   Aero structural Analysis II
Aero Engr. Laboratory   Aerospace Dynamics

4th year   <u>Fall</u>   <u>Spring</u>
Dynamics of Aerospace Vehicles   Aerothermodynamics &
& Propulsion
Aerospace Design Principles   Aerospace Structural Design
Aerospace Systems Design   Fracture Mechanics of
Aerospace
Structures   Aerospace Materials Science

Human Performance in Aerospace

Materials Engineering BS

First year	Fall	Spring
	Chemistry	Physics
	Chemistry Lab	Physics Lab
	Composition	Chemistry
	Engr. Math	Mechanics
	Core Course	

2nd year	Fall	Spring
	Physics	Communication
	Engr. Math	Thermodynamics
	Materials in Society	Soft Matter
	Electricity and Magnetism	Materials Science
	Core Course	Structure of Materials
	Materials Lab	

3rd year	Fall	Spring
	Math for Materials	Properties of Functional Materials
	Materials Lab	Numerical Methods
	Kinetics of Materials	Materials Characterization
	Deformation and Failures	Materials Experiments
	Core Course	Technical Elective

4th year	Fall	Spring
	Materials Design	Materials Design II
	Materials Processing	Core Course
	Core Course	Technical elective
	Technical Elective	Specialty Elective
	Specialty Elective	